DNA Know Thyself

Ray J Rose

DNA Know Thyself

Living in a DNA World

Ray J Rose
School of Environmental & Life Sciences
The University of Newcastle
Callaghan, NSW, Australia

ISBN 978-981-97-8380-9 ISBN 978-981-97-8381-6 (eBook)
https://doi.org/10.1007/978-981-97-8381-6

This Springer imprint is published by the registered company Springer Nature Singapore Pte Ltd.
The registered company address is: 152 Beach Road, #21-01/04 Gateway East, Singapore 189721, Singapore

My late wife Joy Ellen Rose encouraged me to write this book, and it is dedicated to her.

Preface

Although DNA knowledge and its application are extensive, they are often compartmentalized across many disciplines. This book offers a contemporary and integrated view on DNA biology and biotechnology, distinct from a traditional textbook. It is a relatively short read of the key findings leading to contemporary understanding and application. Through our DNA, we are increasingly able to understand ourselves and how our DNA world functions to allow us to live on this planet. Through the knowledge of DNA, it is possible to improve human health, enhance sustainable food production, promote environmental sustainability, and develop novel industries. This book is aimed at the non-specialist who has some background in biology from their education or has a strong interest in DNA biology and biotechnology. While it can provide reading for introductory courses in the medical and life sciences, it is suitable reading for a variety of university degrees as well as other interested audiences. It is also suitable for a stand-alone lecture course. This book can also provide an entrée into specialized DNA and biotechnology areas.

Callaghan, NSW, Australia — Ray J. Rose

Contents

About the Author

Ray J. Rose obtained his bachelor's degree from the University of Sydney and his PhD from Macquarie University, Sydney. He is currently a Professor Emeritus and was a former head of the Department of Biological Sciences at the University of Newcastle, Australia. His research has focused on cell and molecular biology, as well as biotechnology, and he has also taught in these areas. Dr. Rose has published more than 100 peer-reviewed journal articles, including those in *Nature*, *Journal of Cell Science*, *Plant Physiology*, *Proceedings of the Royal Society*, *Series B*, and *Yale Journal of Biology and Medicine*). He has also authored 20 book chapters and edited two books on legume genomics and plant molecular cell biology. He has received an award from the Australian Society of Plant Scientists for his work on chloroplast DNA and division. He was listed as a highly cited and influential author (2004–2008) by the American Society of Plant Biologists, based on an analysis of publications in their two leading international plant science journals, *Plant Physiology* and *The Plant Cell.*

1

Introduction

Humankind has continued to advance its knowledge of planet Earth and of itself since *Homo sapiens* evolved in Africa around 200,000 years ago. Since the discovery of the structure of DNA 70 years ago, knowledge has advanced rapidly. Knowledge from the sequencing of genomes of humans and numerous other species continues to surprise us with the intricacies of how DNA is ultimately responsible for cellular function, inheritance, the diversity of life, and the development and function of every species. Though the DNA knowledge base is extensive, it is compartmentalized into many disciplines. This book seeks to provide a more integrated view of different aspects of DNA biology and biotechnology. The fascinating thing is that through our DNA we can increasingly understand ourselves and how our DNA world functions to allow us to live on this planet. Through the knowledge of DNA, it is possible to work with it to improve human health, sustainable food production, a sustainable environment, and develop novel industries.

The book provides a personal view of DNA in our world and how that knowledge is being used. Knowledge for "Living in a DNA World" is derived from many areas of the biological sciences but which relates back to ourselves, emphasizing the unity of life through DNA. DNA knowledge

R. J. Rose, *DNA Know Thyself*,
https://doi.org/10.1007/978-981-97-8381-6_1

can help us and the planet, but the knowledge needs to be used wisely.

Each chapter gives some historical context on the way understanding has developed, but emphasizes current understanding and provides context for the future. There is a small amount of perspective in relation to contemporary issues. The references in "Further Reading" include overviews, reviews, and significant original papers.

Chapters 2 and 3 consider DNA genesis and the evolution of cells like ours (the eukaryotic cell) and how the diversity of life allows for an environment that sustains us. In Chapters 4 and 5, there is a focus on the explosion of knowledge on DNA and how it is regulated to enable a complex organism like ourselves, with its cellular diversity, to develop from a single cell. The power of stem cells to be programmed into different cells, tissues, and organoids then follows. Chap. 6 deals with the cell cycle of the dividing cell (cell reproduction) and its relation to cancer with its uncontrolled cell proliferation. Our reproduction and inheritance is the focus of Chap. 7. The nature of DNA biology in relation to aging is considered in Chap. 8, a subject which is now a topical research focus. Chap. 9 is devoted to the nature and control of viral infection from DNA and RNA viruses.

The final 6 chapters are focused on the application of DNA knowledge in modern biotechnology. How modern biotechnology developed from DNA cloning in bacteria to what is possible today in microbes, plants, and ourselves is considered in Chap. 10. Chapters 11–13 are devoted to DNA biotechnology in food production and nutrition, the environment and gene therapy for human disease. Chap. 14, which focuses on synthetic biology, has grown out of biotechnology. It uses DNA technology to engineer synthetic cellular systems to produce new products and understanding. The final chapter, Chap. 15, emphasizes the potential of DNA understanding for a better planet and life on Earth if used wisely. Our planet derives its wonder, its ecology, and our lives through a DNA-based world. The knowledge we have should be used to enhance life on Earth.

2

Genesis of DNA and Cells like Ours

We receive our DNA from our parents and pass our DNA to our children to continue human life on our planet. We are also connected through DNA to all creatures that are single cells or composed of many cells. In many ways, it is DNA that makes our planet and each one of us special. In this chapter, how prebiotic chemistry could have contributed to the origin of RNA and DNA is considered as well as potential pathways to primitive cells. At some point, the single cell of the last universal common ancestor (LUCA) gave rise to the three domains of life: the Archaea, prokaryotes, and eukaryotes. It is the eukaryote cells that enabled the diversity of plants and animals and the evolution of ourselves. How did these complex eukaryotic cells evolve? This fascinating story about the endosymbiotic origin of mitochondria and chloroplasts is developed. The origin of the diverse multicellular life forms goes hand in hand with the oxygenation of the atmosphere.

How did this DNA come about? We could go back to the origin of the universe with its beauty and complex physics and chemistry, but I wanted to focus on the planet that we inhabit, which formed 4.56 billion years ago (4.56 Ga). The precise answer to how DNA came to be, like the origin of the universe, is an unknowable, one of those known unknowns. This makes the early chapters of Genesis in the Bible seem comforting. Why worry about mechanisms?

R. J. Rose, *DNA Know Thyself*,
https://doi.org/10.1007/978-981-97-8381-6_2

However, laboratory experiments can be carried out to gain an insight into how the origin of DNA could plausibly have occurred. The details of the most plausible mechanism most likely reside on understanding of how planet Earth formed and an advanced understanding of chemistry and biochemistry.

So, what can we say right now about how DNA came to be? First, we need to dissect DNA a little, whose structure was published by Watson and Crick in the journal *Nature* in 1953. DNA is built up from five elements: carbon (C), nitrogen (N), hydrogen (H), oxygen (0), and phosphorus (P). These five elements are organized into units called nucleotides. A nucleotide in DNA consists of a phosphate group, a 5-carbon sugar called deoxyribose and a nitrogenous base. There are four different nucleotides, each with a different nitrogenous base known as A (adenine), G (guanine), C (cytosine), and T (thymine). The nucleotides are joined together via the phosphate group to form long strands. It is these bases that enable the genetic code made up of sequences of A, G, C, and T. Each one of us has a special sequence of AGC and Ts on each one of our chromosomes. DNA consists of two nucleotide strands joined together in the form of the iconic DNA double helix. The bases enable the classic pairing in DNA structure:

A….pairs with…T.
G….pairs with ...C.

On prebiotic earth, somehow the five elements (C, N, H, O, and P) had to be able to form a string of nucleotides in two strands to produce DNA. At first site, this seems impossibly difficult. However, some simple experiments show that it is possible to form organic molecules (molecules built around carbon) from a cocktail of small precursor molecules that possibly existed at some point on the primitive earth. In 1953, Stanley Miller published an experiment, in which a mixture of methane, ammonia, water, and hydrogen was put into a glass flask and subjected to an electrical discharge to simulate lightning. Though not an author of the Miller paper, published in the journal *Science*, Harold Urey was an important influence on the work and the experiment is sometimes referred to as the Miller–Urey experiment. It is important to contrast the postulated primitive atmosphere with today's atmosphere of carbon dioxide, nitrogen, oxygen, and water. Surprisingly Miller found that amino acids formed in the flask. Amino acids are the building blocks of proteins, which are coded for by DNA via the single-stranded messenger RNA (mRNA) molecule. A more recent understanding suggests that the first atmosphere came from outgassings from the earth's interior, from gaseous volcanic eruptions. The composition was mostly water, carbon dioxide, and nitrogen with small amounts of carbon monoxide and hydrogen, but essentially no methane or ammonia. Nevertheless, the Miller experiment does show that it is possible to produce organic molecules important for life from a mixture of simple precursor molecules that could possibly have existed at some point on primitive earth. Ammonia and methane can come from hydrothermal vents and have been found in meteorites. Importantly no free oxygen was present in the prebiotic atmosphere. But what of DNA? This turns out to be much harder to synthesize from simple components.

There are good reasons to think that RNA preceded DNA in the earliest life and there was an RNA world at one point. The RNA molecule is much more familiar to us today because of mRNA vaccines against COVID-19. Unlike DNA, RNA is single stranded and rather than thymine (T) it contains uracil (U) and instead of the deoxyribose sugar it contains the ribose sugar; otherwise, it is like a single strand of DNA. RNA, like DNA, can store information derived through its four letters (with U instead of T). In addition, we know from current biology that some RNA forms can act like enzymes called ribozymes. This could be important in duplicating an RNA or DNA molecule without the protein enzymes we know today that consist of amino acids.

One way of considering prebiotic RNA synthesis is to approach it in modular terms: form a nitrogenous base, a ribose sugar, a phosphate, then join them altogether to form an RNA strand. Current thinking has hydrogen cyanide (HCN) and its derivatives forming a nitrogenous base. With ribose sugar formation, we need to look at another type of chemical change with formaldehyde (H_2CO) potentially being the progenitor. Importantly, phosphate is present on the earth's surface and bases and sugars can be linked by phosphate (PO_4^-) groups to form the RNA nucleotide strands. It is then a question of understanding the chemistry of how the nucleotides were generated from the nitrogen base, sugar, and phosphate.

If RNA did indeed exist before DNA, there still has to be the formation of DNA from RNA, that is, RNA → DNA. It is now well known from current molecular biology (and the COVID-19 PCR test) that this is possible using a protein enzyme. This was originally discovered in RNA retroviruses (HIV is an example of a retrovirus) that use the reverse transcriptase enzyme to form DNA from RNA. It raises the question as to whether proteins evolved after RNA prior to DNA to enable DNA formation. Alternatively, were RNA ribozymes involved in catalyzing DNA formation before reverse transcriptase-like enzymes ever came to be? Early ideas had prebiotic DNA synthesis occurring in small pools of water. However, a watery environment would favor the breakdown rather than synthesis of DNA molecules. Wet and dry cycles are more likely to enable DNA formation, for example, rain on rocks or clay in the presence of UV radiation from sunlight. Recent photographs from the *Perseverance* rover in 2021 have shown evidence of water forming the rocky environment structure characteristic of the Mars surface. This fuels speculation that an early life-related chemistry on Mars might provide clues for the prebiotic chemistry of nucleic acid formation on earth.

While it is possible that prebiotic chemistry could have contributed to the origin of DNA via an organizational hierarchy, there is no doubt lots of concurrent chemistry occurred during the many millions of years required for the first DNA-containing cells to emerge. A prebiotic pre-RNA evolution and selection for replicating molecules would likely be part of the milieu. For a time, there may have been an RNA world. However, the formation of DNA is critically important given that it is much more stable than RNA and is information rich.

Based on current chemistry and biology, it seems that DNA could potentially form on the prebiotic Earth with its primitive atmosphere and early geology. As indicated in Genesis, we may indeed come from the "dust of the ground." However, a contribution from space cannot be ruled out. Meteors striking earth can have organic molecules and could conceivably have a role in the origin of DNA. Investigation of space will continue to lead to speculation about the key features of the origin of life, based on the chemistries that are revealed.

While it can be argued that DNA is the centerpiece of life processes, it is as part of the cell and its orderly transmission between generations and in the development of multicellular organisms that is critical today. So how did DNA become part of the cell? If earth formed 4.56 Ga years ago, the capacity for the beginnings of life on a habitable earth is estimated at 4.3 Ga and evidence from the rock record of the first microbes, based on morphology and chemistry, is about 3.42 Ga. One of the strongest cases for the first appearance of early life comes from the stromatolites of the Strelley Pool Formation in Western Australia. This gives 0.88 billion years or longer for the first cellular life to evolve: a very long time! To form protocells, the nucleic acid polymers need to be surrounded by a membrane or some type of "wall." The best probiotic candidates are the fatty acids. These can form simple lipid bilayers, albeit much simpler than the membranes that exist today. It is within these protocells that DNA replication could have initiated together with the first protocell divisions. This no doubt was a very long and tortuous process. While DNA molecules surrounded by a membrane might have existed at one point in time in a primitive protocell, there would be a need to develop a biochemical machinery to provide a metabolism that provides energy and the orderly processes of cellular growth and division. The biochemist Jack W. Szostak shared the Nobel Prize for Physiology and Medicine (with Elizabeth H. Blackburn, and Carol W. Greider) for the discovery of "how chromosomes are protected by telomeres and the enzyme telomerase" on the chromosome ends. His current research is on the chemical and physical processes that enable the transition from chemical to biological evolution on early earth. He suggests that fluctuations in the environment, coupled with continued selection, led to the processes of DNA replication, protocell growth, and division. The potential pathways to DNA and protocells on early earth discussed here are backed up by research showing the chemical and physical possibilities of these pathways, but ongoing research continues to put forward alternative chemistries and the type of physical and chemical environment that could have contributed to early life. The chemistry of the universe continues to be unraveled.

One of the arguments directed against the origin of life hypotheses is that thinking is heavily influenced by current knowledge of the biology of the cells that exist today. It is perhaps not surprising given the advances in biology. This is illustrated by the research of Craig Venter, who led the private consortium that sequenced the human genome. Venter's team was able to produce a synthetic or artificial bacterial cell. What they did was to chemically synthesize the total DNA (the genome) of one type of bacteria. The 1.08-mega-base pair genome of *Mycoplasma mycoides was* synthesized: quite a feat. Then, they were able to transplant the *M. mycoides* genome into a recipient cell: *Mycoplasma capricolum*, a different species. The recipient cell did contain the non-DNA part of the cell (cytoplasm and wall) but no DNA. However, what happened was the "dead" chemically synthesized *M. mycoides* DNA coded for all the proteins that were needed to produce another bacterial type, by *M. mycoides* DNA replacing the *M. capricolum* DNA. The Venter experiment does show the power of synthetic chemistry and DNA, and hints at how a cell could assemble.

Nevertheless, there is still a way to go from a protocell to the first cell of the complexity of even the simplest cells known today. Direct evidence for early life is limited, with traces of the first life in early earth sediments and rocks that are both morphological and chemical from biological activity. There is always the possibility that clues might be found from researching exoplanets. At some point, there would be the last universal common ancestor (LUCA), the ancestral one-celled life form that gave rise to the tripartite tree of life, one branch of which gave rise to humankind due to the formation of the eukaryote cell. The tripartite tree, or three domains of life, now based on DNA sequencing, shows how the sequences and type of cell evolved to form the network of life forms.

The acceptance by most biologists of the three domains of life was a major change. Prior to 1977, two domains were widely recognized, the bacteria (prokaryotes) and the eukaryotes. Eukaryotes (we are one of them) have a very distinctive cell structure, the nucleus. The nucleus is bound by a double membrane with its DNA organized into chromosomes. In the nucleus, the genes are transcribed to form RNA, which exits the nucleus through sophisticated nuclear pores. Outside of the nucleus in the cytoplasm are the ribosomes where the RNA is translated to form proteins. The cytoplasm also contains the energy-producing powerhouses, the mitochondria, which also contain a small amount of DNA. In addition to the mitochondria, there are organelles that are built of membranes (e.g., endoplasmic reticulum and Golgi bodies) and others that form a type of skeleton known as the cytoskeleton. Prokaryotes, on the other hand, do not have their DNA in a nucleus and lack organelles such as the endoplasmic reticulum, Golgi bodies, and cytoskeleton. The prokaryotes and eukaryotes have the same genetic code and the DNA → RNA → protein flow, with the RNA translated to proteins on ribosomes, but with some molecular differences.

It was the painstaking work of Carl Woese and colleagues on ribosomal RNA sequences that first led to the discovery of the Archaea (originally called Archaebacteria) as a new domain of life distinct from the bacterial prokaryotes. Ribosomal RNA, together with a number of proteins, are part of the ribosome, the structure involved in protein synthesis. The Archaea domain of life was not readily accepted by microbiologists, but Woese persevered. A quotation by Norman Pace, an evolutionary microbial biologist (commentary in the journal *Science*), puts it this way: “Carl knew he had seen the truth, that he'd seen God in his table of gene relationships." The current understanding is that the three domains of life evolved from a common primordial prokaryotic ancestor and the eukaryotes evolved from the Archaea. As noted, the earliest cells appeared about 3.42 Ga but eukaryotes approximately 1.5 Ga. This is a huge amount of time, almost twice as long as it took to produce the first prokaryotes.

The Archaea, while similar to bacteria, have some unique molecular features plus some molecular features that are similar to eukaryotes. Interestingly, while bacteria are found in almost all environments, Archaea commonly live in extreme environments. These extreme environments include high temperatures (hyperthermophiles), high salt environments (halophiles), or anaerobic (without oxygen) methane-producing (methanogens) Archaea. Methanogens are an interesting group that are particularly relevant today. The methanogens can be found in such places as marine sediments, wetlands, coal seams, and decaying matter, but also in the gut of humans and the rumen of cattle. Methanogens produce methane as a by-product of their metabolism, and cattle are the sources of this gas and contribute to the greenhouse effect. Methane has a lifetime global warming impact 28 times that of carbon dioxide, which has focused attention on this archaeon. In terms of the evolution of the eukaryotes, the closest living relative identified to date is the archaeon informally known as "Loki" of the Asgard group, found in deep marine sediments. These organisms have now been cultured, enabling morphological and cellular characterization.

The first prokaryotic cells to evolve did so in an environment without oxygen, so this had to change if complex multicellular organisms with eukaryotic cells like ours were to evolve. How did this come about? It is a biological phenomenon due to the rise of the prokaryotic cyanobacteria (commonly called blue-green algae) that had the capacity for oxygenic photosynthesis, the ability to produce oxygen (O_2) from water (H_2O) during the photosynthetic process. The cyanobacterial fossil record starts at 3.42 Ga. The Great Oxidation Event, the first rise of oxygen, occurred approximately 2.4 Ga, leading to the oxygen of the atmosphere and oceans. This means that this event took 1 billion years after the evolution of the first prokaryotes (3.42 Ga to 2.4 Ga). Increasing oxygen would facilitate the evolution of the eukaryotic cell (1.5 Ga), which is nearly 2 Ga since the evolution of the first recognizable cell. This emphasizes what a complex event the evolution of the eukaryotic cell was. The eukaryotic cell is the building block of complex multicellular life. The rise of oxygen also led to the formation of the ozone layer (O_3), which protects the Earth from harmful UV (UV-B) rays, damaging to the DNA.

The Archaea of the Asgard group found in fairly extreme environments are now thought of as central to the evolution of the eukaryotes. They contain a number of eukaryotic signature proteins. However, there was a giant evolutionary step that was championed by Lynn Margulis for many years before it became generally accepted. There is now much biological and DNA sequence evidence to support it. It is what is known as the endosymbiotic origin of the DNA containing mitochondria (and chloroplasts) of the eukaryotic cell. Though there had been earlier speculation, Lynn Margulis (then Lynn Sagan) published a 1967 paper where she presented a comprehensive view of the evolutionary origin of the eukaryotic cell from endosymbiotic events. Apparently, this landmark paper had been rejected by a dozen journals before being published. This was the controversial beginning of such an important contribution to biology, perhaps the most innovative evolutionary step in the evolution of the eukaryotic cell. Eukaryotic cells like ours are responsible for the array of multicellular organisms on earth, including humankind.

The development of the mitochondrion by endosymbiosis is a fascinating story. It starts with an archaeon of the Asgard group that engulfed an aerobic bacterium from the aerobic α-proteobacterium class. The Asgard archaeon had cellular protrusions that might trap bacteria, plus a primitive contractile cytoskeleton that could help engulf the bacterium. Over time, the α-proteobacterium became a permanent component of the archaeon cell, evolving into the mitochondrion. These evolutionary changes produced LECA, the last eukaryotic common ancestor, from which all eukaryotes evolved. Much of the DNA of the α-proteobacterium transferred to the nucleus of the archaeon cell and a small but critical amount of DNA became a permanent part of the mitochondria that exists in the eukaryotic cell today. This endosymbiosis theory has strong DNA sequence evidence, and there are types of endosymbioses that still exist today. The mitochondrion, like the cell itself, is able to synthesize DNA and divide to form daughter mitochondria. The evolution of mitochondrion, known as the powerhouse of the cell, contributed to the development of a powerful aerobic metabolism. The mitochondria-containing eukaryote cells are characteristic of fungi and animals including humans. This leaves out another large group of organisms, the plants.

Plants, which include the eukaryotic algae, also have eukaryotic cells with mitochondria. However, they contain another important DNA-containing organelle, the green chloroplast. How did this occur? The DNA sequence, plus biochemical and morphological evidence, also points to an evolutionary endosymbiosis pathway from engulfed cyanobacteria. This endosymbiosis most likely involved a mitochondrion containing eukaryote engulfing a cyanobacterium, occurring approximately 1.2 Ga. The postulated timing is consistent with the earlier protomitochondrion invasion of about 1.5 Ga. The evolution of the photosynthetic eukaryotes would enable a very wide group of organisms to have a photosynthetic capacity. Oxygen levels remained fairly static after the Great Oxygenation Event 2.4 Ga until a second rise of oxygen between 0.8 and 0.4 Ga to near current-day levels of 21%. Interestingly, given endosymbiosis of cyanobacteria occurred 1.2 Ga, photosynthetic eukaryotes could contribute to this second rise of oxygen.

The first eukaryotes were unicellular with an array of different cell biologies: the protists like the amoeba or paramecium and the photosynthetic unicells like *Chlorella*. Coinciding with the increasing oxygen levels began the rise of the multicellular animals and plants leading to the diversity we see today. Based on molecular clock analysis (which uses DNA or protein sequences) and fossil evidence, the origin of metazoan multicellularity occurred about 800 Ma (0.8 Ga) followed by the diversification of animals. This is more than increased cell number, eukaryotic cells can differentiate, becoming specialized for different functions. The DNA positioned in the nucleus has the genetic units (genes) that are regulated to code for the different proteins that are characteristic of the different cell types. The multicellular land plants (the Embryophytes) evolved about 500 Ma before diversifying, originating from their multicellular green Charophyte algal ancestors.

In considering the evolution of more complex animals and plants, the very large times involved and the relatively stable Earth environment compared to other planets known at this time have to be emphasized. The evolution of humankind from animals with complex physiological and anatomical systems again involves long periods of time, the advantages of a planet like Earth and the capacities of the eukaryotic cell. So, in relative time the emergence of humans is quite recent. Fossil remains indicate that modern humans (*Homo sapiens*) were present in Africa about 200,000 years ago. The dating is a changing area dependent on fossil finds and reaching a consensus on when the anatomical features of "modern" humans first appear and could be called the "Adam" and "Eve" of the biblical story of creation. In terms of the age of the Earth, this is so recent. This means that if the Earth is 4.5600 Ga then it took most of that time (4.5598 Ga) for modern humans to evolve. In the DNA and cellular context, this is a triumph of the eukaryotic cell.

It turns out, though, that despite the wonder of the human body and the human intellect it is vulnerable to simpler life forms: predominantly from bacteria and viruses. In the modern world, this is a continuing challenge. But what of the viruses: where do they fit in terms of cellular evolution? Viruses, unlike bacteria, are not cells. In terms of human cells, there are RNA and DNA viruses that hijack the cell's machinery to reproduce themselves before moving on to the next cell and the next host, sometimes being lethal to the host as with COVID-19. COVID-19 is the current plague on humanity in addition to humanity's difficulty in peaceful coexistence and sustaining our ecosystem on an amazing planet.

Further Reading

Betts HC, Puttick MN, Clark JW, Williams TA, Donoghue PC, Pisani D. Integrated genomic and fossil evidence illuminates life's early evolution and eukaryote origin. Nature Ecol & Evol. 2018;2:1556–62.

Cole DB, Mills DB, Erwin DH, Sperling EA, Porter SM, et al. On the co-evolution of surface oxygen levels and animals. Geobiology. 2020;18:260–81.

Dyall SD, Brown MT, Johnson PJ. Ancient invasions: from endosymbionts to organelles. Science. 2004;304:253–7.

Gray MW. Lynn Margulis and the endosymbiont hypothesis: 50 years later. Mol Biol Cell. 2017;28:1285–7.

Javaux EJ. Challenges in evidencing the earliest traces of life. Nature. 2019;572:451–60.

Lanier KA, Williams LD. The origin of life: models and data. J Mol Evol. 2017;84:85–92.

Lazcano A, Bada JL. The 1953 Stanley L. miller experiment: fifty years of prebiotic organic chemistry. Orig Life Evol Biosph. 2003;33:235–42.

Ricardo A, Szostak JW. Origin of life on earth. Sci Am. 2009;301:54–61.

Szostak JW. The origin of life on earth and the design of alternative life forms. Mol Front J. 2017;1:121–31.

Woese CR, Fox GE. Phylogenetic structure of the prokaryotic domain: the primary kingdoms. Proc Natl Acad Sci U S A. 1977;74:5088–90.

3

DNA and the Diversity of Life Sustain Us and Our Environment

The DNA molecule in response to the environment and the associated selection pressures enabled the diversity of life forms that exist on planet Earth. Each species in each of the three domains of life has a unique genetic signature. Occupying the many niches on Earth enables the interactions that enable the sustainability of our ecosystems, our planet, and us. How does the diversity of life contribute? The plants in the photosynthetic process drive the oxygen production of the atmosphere and carbon dioxide capture using the green chloroplasts. Plants and animals use oxygen and produce carbon dioxide in the energy-producing process utilizing mitochondria. Chloroplasts and mitochondria contain small quantities of essential DNA. Our existence is dependent on an atmosphere that provides us with life-giving oxygen and a carbon dioxide balance. Photosynthesis provides sugars containing carbon, hydrogen, and oxygen (CHO) for other carbohydrates and lipids. However, proteins and DNA also require nitrogen and phosphorus. The nitrogen cycle, dependent on bacteria and the Archaea, is outlined. Phosphorus is present in the soil solution and aqueous environment. The carbon cycle, nitrogen cycle, and phosphorus availability have been heavily influenced by humankind.

Initially I wanted to focus on the plants that, together with algae, are photosynthetic eukaryotes. Plants contain chloroplasts and mitochondria that both contain small

R. J. Rose, *DNA Know Thyself*,
https://doi.org/10.1007/978-981-97-8381-6_3

amounts of DNA coding for critical proteins. The chloroplast and the mitochondria are intimately involved in both the Earth's energy dynamics and the gaseous dynamics of the atmosphere. In its green membranes (the thylakoids), the chloroplast has components called photosystems that facilitate the capture of sunlight energy to drive the photosynthetic process of converting carbon dioxide (CO_2) and water (H_2O) to sugars. Sugars consist of carbon, hydrogen, and oxygen (CHO). The process of sunlight energy capture in photosynthesis involves the splitting of water into oxygen (O_2) and hydrogen. Hydrogen consists of an electron (e^-) and a proton (H^+), and it is these electrons and protons that drive the formation of biological energy (ATP-adenosine triphosphate) from sunlight. Little electrical currents are generated in the green chloroplast membranes that, together with the protons, are part of the ATP-producing process that drives the formation of sugars from carbon dioxide and water. The sugars provide C, H, and O for the myriad of molecules in the mature plant, which are in turn consumed by animals and ourselves.

The oxygen produced in the water-splitting process is essential for the O_2 used for ATP production in the mitochondria of plants and animals in the respiration process. Oxygen is also used in maintaining the ozone layer (O_3) and the 21% oxygen levels of the atmosphere. The ozone layer protects the earth from harmful UV-B rays.

A key enzyme in the utilization or fixation of carbon dioxide in the formation of sugars is Rubisco (ribulose bisphosphate carboxylase-oxygenase), the carbon dioxide fixation enzyme. It is the biological energy from sunlight, driving a cycle known as the Calvin cycle, that allows this carbon dioxide fixation by Rubisco. The Calvin cycle is named after Melvin Calvin, who received the Nobel Prize in Chemistry in 1961. Melvin Calvin, Andrew Benson, and James

Bassham were co-discoverers of the cycle at the University of California, Berkeley in the 1950s. The Rubisco carbon dioxide fixation, in an ecologically sustainable environment, maintains the carbon dioxide levels that provide an atmosphere like the one that existed before human-induced carbon dioxide emissions.

DNA outside the nucleus is critical for the formation of Rubisco. This DNA is in the chloroplast. A gene in chloroplast DNA codes for a subunit of Rubisco (called the large subunit) while a gene in the nucleus codes for another subunit (called the small subunit). The small subunit is formed in the cytoplasm of the cell and is transported to the chloroplast, which with the large subunit is assembled into a functional Rubisco. This arrangement reflects on the way chloroplasts evolved from the cyanobacteria endosymbiosis.

The mitochondrion is the other organelle outside the nucleus that contains a small amount of DNA that is critical in maintaining the living plant with its photosynthetic capacity. What the mitochondria ultimately do is produce ATP in the process known as respiration by metabolizing sugars produced by photosynthesis. As with photosynthesis, both electrons (e^-) and protons (H^+) are involved. In the case of mitochondria, the electrons flow in the cristae of the inner mitochondrial membrane. This biological energy is then used to enable plant growth and development using sugars and other compounds that provide nitrogen and phosphate for proteins and nucleic acids. In respiration (in plants and animals), to enable the energy formation, oxygen is utilized and there is an associated cycle in the mitochondria where carbon dioxide is produced. This latter cycle is widely known as the Krebs cycle (or citric acid cycle). Hans Krebs received the Nobel Prize in 1953. Plants provide life-giving oxygen and in an environmentally balanced world utilize carbon dioxide to produce sugars from photosynthesis for their growth, maintaining, or producing new photosynthetic capacity.

What is critical in terms of the genes involved in aerobic respiration is that some critical genes are present in mitochondrial DNA and some in the nucleus. It is still somewhat of a puzzle as to why a cell would have critical genes located in two compartments. Evolution via endosymbiosis has dictated the way mitochondrial and nuclear genes must work together. The consequence of having chloroplasts and mitochondria with their special structure and their DNA is that the maintenance of these organelles and their accompanying DNA is essential for sustaining life as we know it. Like cells, chloroplasts and mitochondria grow and divide and transmit their DNA to their daughter chloroplasts and mitochondria. All chloroplasts and mitochondria are derived from preexisting chloroplasts and mitochondria. Good-quality chloroplasts and mitochondria need to be maintained in the plant cell. Damaged chloroplasts and mitochondria can be removed in a process known as autophagy. Chloroplasts and mitochondria are transmitted across generations. Interestingly, in most cases chloroplasts and mitochondria are transmitted across generations by the female parent.

The cyanobacteria are the prokaryotes that are able to carry out oxygenic photosynthesis, and as discussed in Chap. 2 are the evolutionary precursors of the chloroplasts of the eukaryote algae and plants. Cyanobacteria are an important part of the marine environment and, together with algae, are part of the plankton that underpins the marine ecosystem.

Our existence is dependent on plants to enable an atmosphere that provides us with life-giving oxygen, nutrition, and a carbon dioxide balance for an optimum environment. However, the system is starting to get out of balance because we are pumping unsustainable levels of carbon dioxide into the atmosphere and eliminating trees and other forms of plant life. Carbon dioxide in excess of pre-industrial levels leads to global warming and climate change. Pre-industrial CO_2 levels were about 280 ppm but are now about 416 ppm due to anthropogenic carbon dioxide. Carbon dioxide is important for life and maintaining an atmosphere that provides livable temperatures. If atmospheric CO_2 gets too high, it forms what amounts to a suffocating blanket trapping the heat from the earth's surface—the greenhouse effect. Ironically what we see in nature provides the conceptual solution to climate change. Like in photosynthesis, sunlight can be used to produce electrons for our power usage. There is also increasing interest in hydrogen, produced by the hydrolysis of water as a fuel source. It is interesting that we don't have to look beyond nature to solve our energy problems—sunlight and water (plus human ingenuity) are all you need!

Plants, like us, require more than carbohydrates such as sugar for their growth, development, and reproduction. Fats and oils like carbohydrates contain CHO but in different molecular structures. But, what of proteins and DNA? Proteins also contain nitrogen (N) and DNA also contains phosphorus (P) and nitrogen. Where do plants get nitrogen and phosphorus from?

When N is considered, another group of organisms come to the fore—the bacteria and Archaea. These microbes are an essential part of the nitrogen cycle supplying plants with a form of nitrogen they can use. Nitrogen (as N_2) is 78% of the atmosphere but it is not in a form that can be utilized by plants. Most plants must get their nitrogen from the soil as nitrate (NO_3^-), or less commonly as ammonia (NH_3) taken up from the soil as ammonium ions (NH_4^+). How does this come about? There are diverse forms of prokaryotes (and some archaeons) that are able to "fix" atmospheric nitrogen; all using an enzyme called nitrogenase to form ammonia. Nitrogenase is very sensitive to oxygen, so there must be strategies for aerobic microbes to protect the enzyme from oxygen. The ammonia can be converted to nitrate by nitrification, first involving nitrite formation then nitrate formation. There is a group of aerobic bacteria that can convert ammonia into nitrite (NO_2^-). Ammonia-oxidizing archaeon have also been discovered. A separate group of bacteria, known as nitrite-oxidizing bacteria, can convert nitrite to nitrate (NO_3^-) that can be readily used by plants. The two-step process for nitrification involving two different groups of bacteria was classical understanding for more than 100 years. In 2015, it was discovered that the two-step nitrification process could be carried out by a single organism in bacterial species known as comammox bacteria (complete ammonia oxidizers). The *Nitrospira* comammox species encode all the necessary enzymes for ammonia oxidation to nitrite and nitrite oxidation to nitrate.

While atmospheric nitrogen can be converted by biological processes to nitrate, the nitrogen in nitrate can be returned to the atmosphere as nitrogen gas to complete the nitrogen cycle. Again, these denitrification processes are carried out by a different group of bacteria commonly under anaerobic conditions but can occur under aerobic conditions. The different types of bacteria and archaeons that participate in the nitrogen cycle have specific enzymes coded for by DNA that enable the nitrogen cycle to occur in land and aquatic environments, with different prokaryotes and archaeons characteristic of the environment. Though nitrogen gas (N_2) is the ultimate product of denitrification, other intermediate gaseous forms of nitrogen are produced in the process. One such form is nitrous oxide (N_2O), which is a greenhouse gas.

The nitrogen cycle, like the carbon cycle, has been heavily influenced by humankind. As nitrogen is so important to plant growth, it is often supplied as nitrogenous fertilizer to agricultural crops. If there is excessive use, nitrate can leach from the soil into river systems, or as in Australia into the Great Barrier Reef, where it can lead to algal blooms and ecosystem destruction. Nitrogenous fertilizers can increase N_2O and contribute to the greenhouse effect. As a greenhouse gas, nitrogenous oxide is about 300 times more potent than carbon dioxide. It has been estimated that N_2O from synthetic nitrogenous fertilizers contributes about 2.1% of global greenhouse emissions.

In addition to the free-living microbes supplying soil nitrate for plant growth, there is a group of plants known as legumes that can access atmospheric nitrogen in another way. Legumes include agricultural plants such as soybean, chickpea, and alfalfa (lucerne), as well as being present in natural ecosystems. Legumes have a symbiotic association in their roots with a group of bacteria known as rhizobia. These bacteria, housed in nodules, contain the nitrogen-fixing enzyme nitrogenase that produces ammonia, which can feed into amino acids. This is an excellent arrangement, with the plant supplying photosynthetic carbon products and the rhizobia contributing available nitrogen that enables amino acid biosynthesis. The use of legumes in agricultural systems can reduce the adverse effects of nitrogen fertilizer by reducing run-off of nitrate and nitrogenous oxide emissions. Legumes clearly have an important role to play in ecosystems and cropping programs in agriculture. It is also noteworthy that nitrogenous fertilizers are produced by the Haber–Bosch process, a very energy-intensive process, accounting for about 1% of all energy use.

Plants can use carbon dioxide from the atmosphere, atmospheric nitrogen converted to an available form and water to produce carbohydrates, lipids (fats and oils), and proteins utilized in the human diet either directly from plants or via the food chain. Humans cannot synthesize certain amino acids known as the essential amino acids. These nine essential amino acids can be supplied by plants. For example, soybean, a legume, is a source of all nine essential amino acids.

So far we have considered how plants can obtain C, H, O, and N. But what of phosphorus, present in DNA nucleotides, in phospholipids of cell membranes and the key energy molecule ATP? Where does phosphorus come from? Phosphate salts in rocks and other mineral deposits are weathered so that inorganic phosphate (Pi) exists in the aqueous environment or soil solution and is taken up by the plant as a phosphate (H_2 PO_4^- or HPO_4^{2-} depending on pH).

Phosphorus can be quite limiting for agricultural crops; maintaining a constant supply for the roots to utilize is difficult. The plant itself can mine more Pi by having an expansive root system. But here again other organisms help. There are fungi that have a mutualistic or symbiotic association with roots of many species. They form what is known as the arbuscular mycorrhizal (AM) symbiosis that colonizes the root cortex and the mycelium reaches out into the soil where the hyphae spread throughout the soil-promoting nutrient uptake into the plant, particularly Pi. There is also a group of bacteria that can hydrolyze organic and inorganic phosphorus to soluble Pi that can be taken up and utilized by the plant. Phosphorus is such an important nutrient for high plant yields, and given it may be limiting in many soils, it is not surprising that phosphate fertilizers are widely utilized. Phosphate fertilizers such as superphosphate are based on rock phosphate, which is mined and is a nonrenewable resource. It is not widely distributed, and there are variable estimates of when it will be exhausted. Global reserves may last 300–400 years but with reserves concentrated in a few countries. It would be wise to focus more on recycling organic residues and maintaining a sustainable phosphate cycle.

The downside of phosphate fertilizers is that they tend to be over-utilized and like nitrogen fertilizers can cause algal blooms (algae and cyanobacteria, which are sometimes called blue-green algae) due to run-off into streams, lakes, and oceans. There is decreased dissolved oxygen because of the algal blooms with their bacterial decomposition causing eutrophication. The algal blooms also produce toxins that can be harmful to human health. One of the scandals of phosphate mining is the mining of the Pacific Island of Nauru. The world's purest deposit of rock phosphate that once blanketed the island has been almost completely removed exposing a buried landscape of karst pinnacles. An extreme example of human-induced environmental degradation.

To sustain life, including our own, there are these natural cycles dependent on specific organisms with their characteristic DNA code that influence our environment and the supply of the major molecules for our growth and development. The plants and other photosynthetic organisms supply the oxygen we breathe and the sugar ($C_6H_{12}O_6$) that supplies the carbon, hydrogen, and oxygen utilized in life's biochemistry and the gaseous balance of our precious atmosphere. Specific bacteria and archaeons (and leguminous plants) enable the atmospheric nitrogen (N_2) to be ultimately utilized in the amino acids of our proteins. Fungi and certain bacteria contribute to the phosphorus (Pi) cycle. Phosphate is an important part of our DNA and in ATP is an important part of our biological energy systems. We must protect the organisms and the cycles they are involved in and reimagine best practice to sustain them, which needs bioscience, agriculture, industry, and economics to work together.

Further Reading

Bernhard A. The nitrogen cycle: processes, players, and human impact. Nat Educ Knowl. 2010;2:12.

Ferrol N, Azcón-Aguilar C, Pérez-Tienda J. Arbuscular mycorrhizas as key players in sustainable plant phosphorus acquisition: an overview on the mechanisms involved. Plant Sci. 2019;280:441–7.

Fischer WW, Hemp J, Johnson JE. Evolution of oxygenic photosynthesis. Annu Rev Earth Planet Sci. 2016;44:647–83.

Johnson MP. An overview of photosynthesis. Essays Biochem. 2016;60:255–73.

Kalayu G. Phosphate solubilizing microorganisms: promising approach as biofertilizers. Int J Agron. 2019;2019:4917256.

Mokhele B, Zhan X, Yang G, Zhang X. Nitrogen assimilation in crop plants and its affecting factors. Can J Plant Sci. 2012;92:399–405.

Rose RJ. Sustaining life: maintaining chloroplasts and mitochondria and their genomes in plants. Yale J Biol Med. 2019;92:499.

Schwember AR, Schulze J, Del Pozo A, Cabeza RA. Regulation of symbiotic nitrogen fixation in legume root nodules. Plants. 2019;8:333.

Van Kessel MA, Speth DR, Albertsen M, Nielsen PH, Op den Camp HJ, et al. Complete nitrification by a single microorganism. Nature. 2015;528:555–9.

Zhang H, Zhao Z, Chen S, Kang P, Wang Y, et al. *Paracoccus versutus* KS293 adaptation to aerobic and anaerobic denitrification: insights from nitrogen removal, functional gene abundance, and proteomic profiling analysis. Bioresour Technol. 2018;260:321–8.

4

The Evolving Story of DNA, Genes, and Genomes

Last year, 2023, was the 70th anniversary of the discovery of the structure of DNA by Watson and Crick. In those 70 years, there has been an amazing growth in the knowledge of the biology of DNA; as chemistry, biochemistry, and molecular biology, cell biology, genetics, physics, and bioinformatics have been brought to bear. The central dogma of DNA → mRNA →protein has been refined. Restriction enzymes for cutting DNA at specific sites led to recombinant DNA and the cloning of individual genes. The interrupted eukaryotic gene is organized into exons and introns, with the introns being spliced out. Discovery of DNA sequencing led to the human genome project. This was followed by new-generation sequencing technology and the explosion of genome sequencing. The human genome sequencing has revealed that the human genome has much more than individual protein-coding genes. Noncoding RNA genes have been revealed. The role of nucleosomes in the organization of the genome has been refined. Much DNA in the nucleus is devoted to how genes are regulated leading to the development of human complexity.

In the 70 years since the discovery of DNA structure, there has been an amazing growth in the knowledge of the biology of DNA. Alongside the biology, the engineering of instrumentation and advances in computing have also played an important role. Several years prior to the

R. J. Rose, *DNA Know Thyself*,
https://doi.org/10.1007/978-981-97-8381-6_4

discovery of the DNA structure, it was shown that DNA was the genetic material by using bacteria and bacteriophages (viruses that infect bacteria). Many years prior to that, Gregor Johann Mendel, a monk in a monastery, worked with peas in the monastery garden to show traits were inherited and that there were units of inheritance. His work *Experiments on Plant Hybrids* was published in 1866 in *The Proceedings of the Brün Society of Natural History*, and its significance rediscovered in 1900. Similar experiments can be readily done with laboratory model eukaryotes used in genetics, such as the mouse, the fruit fly *Drosophila,* and the model plant *Arabidopsis.* Starting with the Last Universal Common Ancestor (LUCA), eukaryote cells ultimately led to the multicellular animals and, finally, to *Homo sapiens.* Human beings then unraveled the very molecule that encoded the information that defined our own human potential and our connection to all life forms.

Knowing the structure of DNA led to the formulation of the central dogma for the flow of genetic information. First conceived by Francis Crick in 1957, it indicates information from DNA flows to messenger RNA (mRNA), which in turn specifies the protein.

DNA → mRNA →Protein.

While the central dogma had to be refined, it remains fundamental to molecular biology. Somehow RNA, with its four-letter code due to the four different bases (AGCU), had to specify the sequence of the 20 amino acids that are present in proteins. In the 1960s, it was shown that three bases, the triplet codon, specified a particular amino acid. Marshall Nirenberg, Har Gobind Khorana, and Robert Holley were awarded the Nobel Prize in 1968 in recognition of their roles in the deciphering of the genetic code. This was due to the synthesis of RNA oligonucleotides (short strings of nucleotides) and then testing their ability in a test tube to bind amino acids to the ribosome protein-synthesizing machinery. RNA is the message from the sequence of a specific gene that specifies a particular protein, a chain of amino acids, a polypeptide.

A gene then is a stretch of DNA coding for a single polypeptide. While this still holds true, it is now known that this needs elaboration. While an enzyme (RNA polymerase) reads the DNA to synthesize mRNA, there is an enzyme (reverse transcriptase) that can synthesize DNA from mRNA. Some genes do not code for a protein but encode regulatory or structural RNAs. In non-eukaryotes such as bacteria, the mRNA reads a linear sequence of amino acids present in the polypeptide. However, in 1977, somewhat surprisingly, it was shown that most eukaryote genes are interrupted or "split." Richard Roberts and Phillip Sharp showed that the eukaryote gene consisted of exons and introns. The exons are the coding pieces, and the introns are noncoding and must be spliced out, leaving the exons producing a linear coding sequence. Roberts and Sharp were awarded the 1993 Nobel Prize for Physiology and Medicine. This turns out to be very important, particularly in humans where around 60% of genes may be alternatively spliced, that is, splicing occurs at different points, so one gene has the potential to form more than one polypeptide. Eukaryotes have a membrane-bound nucleus where the gene is transcribed to form pre-mRNA and is spliced in the nucleus to remove introns to form a mature mRNA. The mature mRNA migrates to the cytoplasm where it is translated to a polypeptide on the ribosome.

In addition to the discovery of the interrupted eukaryote gene, there were other important discoveries in the 1970s that accelerated DNA research. Due to restriction enzymes that cut DNA at specific sites and produce "sticky ends," the formation of recombinant DNA molecules from DNA of diverse organisms was possible. Paul Berg produced the first recombinant DNA molecules involving different organisms in 1972. Using the circular DNA tumor virus SV40, the viral DNA was cut and a DNA fragment containing three genes from the DNA of *Escherichia coli* inserted into the SV40 virus. Subsequently, Stanley Coen and Richard Boyer in 1973 showed that individual genes could be cloned in bacterial plasmids to produce multiple copies for analysis. Plasmids are circular pieces of DNA that are carried along with but separate to the bacterial genome and are also replicated during bacterial growth and division. Researchers were very concerned about the potential for risky experiments involving recombinant DNA, and in February 1975 a conference was held at Asilomar, California. This conference influenced the development of containment and safety procedures related to the degree of risk with the genes involved. Also, in the 1970s, DNA sequencing was developed by Walter Gilbert and Frederick Sanger and was critical in furthering DNA research in humans and the biological world. The Sanger method became widely used. Molecular biology became a rich vein for awarding Nobels. In 1980, Berg, Gilbert, and Sanger were awarded the Nobel Prize in Chemistry. The sequencing of a bacterial virus (phi-X_{174}) of 5375 base pairs by Sanger and coworkers in 1977 led to sequencing of the genomes of free-living organisms. The first bacteria, and first free-living organism, to be sequenced was *Haemophilus influenzae* in 1995 by a team led by Craig Venter. This disease-causing organism has a genome of 1,830,137 base pairs. This organism can cause bacterial infections, especially in children, causing diseases such as meningitis and pneumonia. Fortunately, a vaccine has dramatically reduced these infections. The complete sequence of *E. coli* of 4,639,221 base pairs, widely used in DNA cloning and molecular biology, was published in

1997. In humans, the sequence of the small mitochondrial genome, consisting of 16,569 base pairs, was published in 1981.

The development of Sanger DNA sequencing methods and subsequently automated sequencing machines meant that the sequencing of the human genome could be initiated. In this case, 3.2 billion sequences had to be sequenced. The publicly funded international Human Genome Project took a decade to be completed and was led by Francis Collins, who was appointed head of the National Center for Human Genome Research (NCHGR), a US agency. James Watson, co-discoverer of the DNA double helix structure, was instrumental in enabling funding to set up the NCHGR and headed the project (started in 1990) until April 1992. The public human genome sequence was published in February 2001 in the journal *Nature*. The public international project was based on a clone-based map of the human genome. Segments of about 150,000 base pairs were inserted into bacterial artificial chromosomes (BACs) and inserted into bacteria and cloned. A fingerprint of each BAC (using restriction enzymes that cut at specific sequences) allowed overlaps between BACs to be determined. The BACs could then be lined up to produce a physical map of the genome. Each BAC could be sequenced at any one of a number of international labs participating in the project; mainly in US labs such as the Whitehead Institute located in the Massachusetts Institute of Technology (MIT), the Sanger Centre in the United Kingdom, and labs in Japan, Germany, France, and China.

Another project to sequence the genome using another strategy was carried out by a private company, Celera Genomics of Rockville, Maryland, headed by Craig Venter. This project took only 3 years using a shot-gun sequencing approach. This approach was first used in the sequencing of the bacterium *H. influenzae* by Venter to eliminate the need for initial mapping. Random fragments are sequenced and their relationships determined using computer power to find overlaps and order the fragments. Assembly was no doubt helped by the mapping approach of the public project. The Celera sequence was also published in 2001 in the journal *Science*. The human genome sequence was announced at a joint press conference by Venter and Collins. The first two sequences of the human genome were essentially an initial draft and have been continually refined. In 2022, the human genome sequence (but not the Y chromosome), regarded as accurate and complete, was published. The complete sequence of the human Y chromosome was published in 2023. Continued improvements in technology allowed all the gaps in the difficult repetitive regions to be filled in.

In parallel with molecular biology advances, computer databases are necessary to assist in understanding DNA. All sequencing information, which grows daily, can be entered in the NCBI, the US-based National Centre for Biotechnology Information, part of the United States National Library of Medicine. This is a very important resource, and arguably the most important, to assist human health research, the understanding of life processes, and develop biotechnology. Improvements in computing power enable many genes to be explored, relationships between organisms, human genetics, gene regulation, molecular clocks and evolution, and fighting disease across the biosphere. At the time the human genome was reported, model eukaryote organisms had also been sequenced. Model organisms have been selected to aid in understanding the function of individual genes, known as functional genomics. Important model eukaryote organisms are the single-celled *Saccharomyces cerevisiae* (yeast),

Caenorhabditis elegans (transparent multicellular nematode worm), *Drosophila melanogaster* (fruit fly), and *Arabidopsis thaliana* (thale cress, a tiny flowering plant). *S. cerevisiae* has been widely used in understanding the cell division cycle and linking that to cancer. *C. elegans* is used in understanding neural development and aging. *D. melanogaster* has contributed greatly to basic genetics and how genes are inherited in populations. *A. thaliana* is helping understand how plants work and contributes to understanding the biosphere and human nutrition. What is useful with models is that genes with similar functions are often conserved across nature or provide clues to the function of human genes.

Since the human genome project completion, there have been major changes in DNA sequencing. To sequence the 3.2 billion nucleotides that make up the human genome took a decade and cost over a billion dollars. Next-generation DNA sequencing technology (NGS) in 2022 could sequence a human genome in 1 day for a cost of about $1000 dollars. With this technology, DNA is fragmented into many small pieces and there is massive parallel sequencing of these small sequences. Each fragment is decoded on a chip using fluorescently colored nucleotides added by an enzyme. Computing power then assembles the genome. NGS has led to massive changes in genomics. While models are still important in functional genomics, many types of genomes have been sequenced to illuminate understanding of evolution and human genes. The sequence of the human genome is now more accurate and can be used in the genetics of disease phenotypes. An improved understanding of human genetic diversity has come from the sequencing of large numbers of diverse human genomes. Genome sequencing of COVID-19 is used to monitor variants and determine vaccination strategies.

Once whole genomes could be sequenced, estimates of the number and types of genes could be determined, with information sharing between genomes helping the process. Some genes were well known from detailed studies of their biochemistry and molecular biology. However, there were and are many genes of unknown function. One immediate question after genome sequencing was how many protein genes in the genome are there? This question has been hard to answer precisely. Before the human genome was sequenced, there were estimates of more than 100,000. At the completion of the human genome project, estimates were around 35,000 and with more investigation of individual genes (as opposed to predictions) estimates are now about 20,000. The individual human genes are now in databases with the exact count varying as individual genes are investigated in more detail. What is apparent from comparisons with other organisms is that protein gene number does not in itself account for complexity, for example, rice estimates are about 40,000 based on genome sequencing. Rather than gene number, complexity is due to the way genes interact, that is, regulated. Protein-coding exons in the human genome account for approximately 1% of the genome. While there are 20,000 genes in the human genome, there are more proteins than 20,000. Just how many is still not clear. About two-thirds of the genes in the human genome are split genes (i.e., with introns), so with alternate splicing where not all exons are included in the protein, there is the potential for additional proteins, beyond 20,000. The total number of proteins of an organism is called the proteome. The human proteome may be as high as 50–60,000. Given that there is only approximately 1% of the genome (the exons) coding for proteins, what is the rest of the genome doing? To start with, exons only account for a small proportion of protein-coding genes, with 24% of the genome being introns as opposed to the 1% of exons.

If approximately 25% of the human genome is devoted to protein-coding genes, what are the remaining sequences doing? It was initially thought that much of this was essentially junk DNA accumulated during evolution, perhaps having a structural role but not a functional role. This is a similar situation for most eukaryotic genomes. However, with the ability to sequence all cellular RNAs (RNAseq) it is now known that most of the human genome is transcribed, it is a veritable RNA machine. What is now clear is that not all RNAs are translated into proteins via mRNA. These are the ncRNAs (noncoding RNAs). The ncRNA genes produce RNAs that regulate and coordinate the protein-coding genes. There are different types of ncRNAs with different regulatory functions. Some of these RNA genes like the primary-microRNAs (pri-miRNAs), which are precursors of miRNAs, can be in introns. The miRNAs are processed from pri-miRNA genes into double-stranded miRNAs, which set in train mechanisms that can inhibit mRNA translation—a form of gene regulation. The miRNAs, which are only 22 nucleotides in length, are one of the small noncoding RNAs (sncRNAs). There is another group of RNA genes known as long ncRNAs (lncRNAs), which are the size of messenger RNAs and can also have a regulatory function. There may be large numbers of different uncharacterized lncRNAs in low copy number involved in organizing the genome and shutting down transcriptionally inactive regions. Just how many small and long RNAs there are is yet to be fully determined. Estimates of lncRNAs are in the order of 17,000–20,000 and 10,000 for sncRNAs. This is a work in progress. In any event, only a small proportion of the genome is involved in producing RNAs coding for proteins.

All the DNA sequences present in a genome are not simply like a long length of string in the nucleus. The genome is subdivided and packaged into discrete chromosomes. Our cells contain 46 chromosomes, 23 from each parent. Chromosomes are visible during mitosis of the dividing cell after they have replicated in interphase, with a copy of each chromosome passed on to the two daughter cells. Some of the DNA sequences are involved in forming the chromosome and account for some more of the DNA not involved in protein synthesis or gene regulation. The centromere is the region of the chromosome where the spindle microtubules attach to partition the duplicated chromosomes to the new cells. Centromeric DNA consists of highly repeated sequences. The ends of chromosomes also contain highly repeated sequences known as telomeres and are critical to the integrity of the chromosome. The chromosomes in the new cell decondense or unwind to fit into a newly formed interphase nucleus.

In the nucleus, the chromosomes must be packaged, which presents a problem but leads to further understanding about the genes we have. The total length of the DNA in the human nucleus drawn out like a string would be 2 meters. How does all that DNA fit into the microscopic nucleus? In the interphase nucleus, DNA does not occur as a linear continuous strand but is organized like beads (nucleosomes) on a string. Imagine a string of beads. Each bead consists of special proteins known as histone proteins. The DNA is wound around a bead, then there is a linking piece of DNA before winding around another bead until the end of the chromosome. This beaded structure allows the packaging of each chromosome into the nucleus, where each chromosome has its specific territory. Looking again at the gene means re-evaluating what is required to transcribe mRNA when DNA is wrapped around nucleosomes. This is important in understanding how each gene is regulated. How each cell type has its own set of genes switched on. Switching on a gene requires action on nucleosomes, enabling the enzyme (RNA polymerase) that reads the DNA to form RNA to be set up and activated at the promoter adjacent to the start of the gene. This will be considered in Chap. 5.

Further Reading

Amaral PP, Dinger ME, Mercer TR, Mattick JS. The eukaryotic genome as an RNA machine. Science. 2008;319:1787–9.

Amaral P, Carbonell-Sala S, De La Vega FM, Faial T, et al. The status of the human gene catalogue. Nature. 2023;622:41–7.

Berk AJ, Sharp PA. Sizing and mapping of early adenovirus mRNAs by gel electrophoresis of S1 endonuclease-digested hybrids. Cell. 1977;12:721–32.

Cobb M. 60 years ago, Francis Crick changed the logic of biology. PLoS Biol. 2017;15:e2003243.

Fleischmann RD, Adams MD, White O, Clayton RA, et al. Whole-genome random sequencing and assembly of *Haemophilus influenzae* Rd. Science. 1995;269:496–512.

Goodwin S, McPherson JD, McCombie WR. Coming of age: ten years of next-generation sequencing technologies. Nat Rev Genet. 2016;17:333–51.

Kornberg RD, Lorch Y. Twenty-five years of the nucleosome, fundamental particle of the eukaryote chromosome. Cell. 1999;98:285–94.

Nirenberg M, Leder P. RNA codewords and protein synthesis: the effect of trinucleotides upon the binding of sRNA to ribosomes. Science. 1964;145:1399–407.

Walker J. Frederick sanger (1918–2013) double Nobel-prizewinning genomics pioneer. Nature. 2014;505:27.

Watson J, Crick F. Molecular structure of nucleic acids: a structure for deoxyribose nucleic acid. Nature. 1953;171:737–8.

5

We Come from Single Cells: From Gene Regulation to Stem Cells

The DNA from both of our parents comes together in the fertilized egg cell, the zygote. This single cell can produce the adult Homo sapiens with all the different cell types. This is quite amazing, as amazing as the evolutionary history of Homo sapiens. The process of cell division means each new cell has the same DNA content and yet many different cell types can be faithfully produced at the right place at the right time. Different cells have different genes switched on and off. Transcription of the gene is regulated at the promoter region. Transcription in eukaryotes is more complex because of the presence of nucleosomes. Small noncoding RNAs can be involved in epigenetic regulation of transcription. In adult humans, different types of cells can be derived from stem cells. An example of stem cells in adult humans is the pluripotent cells in the bone marrow producing different blood cell types. Pluripotent stem cells also occur in plant meristems. Embryonic stem cells as well as induced pluripotent stem cells from mature cells are used for in vitro pluripotency. Switching stem cells into different cell types is considered as well as stem cell-based regenerative medicine, organoids, and embryo models.

We come from a fertilized egg cell, the zygote, containing the DNA of both parents. Though it should not be forgotten that the zygote also has the cytoplasm with its organelles, particularly the mitochondria, it is the DNA in the

R. J. Rose, *DNA Know Thyself*,
https://doi.org/10.1007/978-981-97-8381-6_5

nucleus that is the major player. Different cell types in an organ have historically been based on structure and function. Though each cell type has the same DNA, different genes are switched on or off. Each cell type has a different set of expressed genes that can be shown by RNA analysis. Contemporary studies have been based on the ability to obtain gene expression and protein data from individual cells. Thousands of cells have been analyzed to form cell atlases of human organs. More cell types and subtypes have been defined, providing new data for understanding organ function in normal and disease states. How do the different cell types come about?

To be active, an individual gene must be transcribed to produce RNA. The gene contains the sequence of nucleotides for a specific mRNA (messenger RNA) that codes for a specific protein. Adjacent to the gene's start sequence, upstream, is the promoter sequence. It is what happens at the promoter that determines whether the gene is transcribed to produce mRNA. Transcription is carried out by an enzyme called RNA polymerase II (RNA Pol II), which must bind to the promoter and be activated to start reading off the mRNA from the DNA. The binding and activation of RNA Pol II requires three types of proteins: general transcription factors, gene-specific transcription factors, and mediators. The general transcription factors form the pre-initiation complex (PIC), which allows RNA Pol II to recognize and bind to the promoter. To turn on the correct gene required for that cell type, gene-specific transcription factors must recognize DNA sequences that are at adjacent sites upstream from where RNA Pol II is bound or further upstream enhancers. The gene-specific transcription factors recruit mediator, which activates RNA Pol II associated with the PIC to allow transcription to begin. For the mediator to make this connection, some bending or looping of the DNA is involved. It should be noted that the mediator and the PIC consist of a number of proteins forming a complex. From this mechanism, it can be seen how gene-specific transcription factors can lead to regulating groups of genes by having genes requiring the same specific transcription factor. Also, different cell types can utilize a defined set of specific transcription factors.

What has been considered so far makes it seem that we are dealing with a long strand of DNA uninterrupted by nucleosomes that occur along the DNA. So, what happens to the nucleosomes? The promoter region sequences where the RNA polymerase binds are nucleosome free, but the transcribed region contains nucleosomes. In terms of nucleosome-free promoter regions, nucleosomes are not necessarily uniformly positioned along the DNA. There are nucleosome-remodeling enzymes capable of repositioning nucleosomes such that there are nucleosome-depleted regions. In the transcribed regions, the RNA Pol II essentially transcribes through the nucleosomes, with concomitant modification of the nucleosomes. The nucleosomes have DNA wrapped around octamers of histone proteins, consisting of two copies each of H2A, H2B, H3, and H4. The histone H1 is the linker histone, connecting the nucleosome to the linker DNA between the nucleosomes. Histones can be modified by several processes (acetylation, methylation, ubiquitination, and phosphorylation) and transcription elongation factors that influence the DNA–histone interactions.

Transcription initiation and the formation of the RNA transcript in eukaryotes are more complex because of the presence of nucleosomes. While the role of nucleosomes in preventing the 2 meters of DNA in the nucleus from getting tangled up is important, Roger Kornberg, a Nobel Prize winner in 2006 for his work on the transcription apparatus in mRNA formation in eukaryotes, has stressed another role. Kornberg was also instrumental in elucidating nucleosome structure. Together with his wife, Yahli Lorch, they suggested nucleosomes evolved primarily to regulate transcription in eukaryotes, especially for preventing transcription. As indicated above, RNA Pol II is set up at the promoter by a complex machinery, at nucleosome-free sequences. The presence of nucleosomes would prevent this and genes would be inactive. There is an advantage in blocking transcription and having genes turned off, with only the active genes suitable for a particular cell type. Nucleosomes are also important for packaging the DNA into the compacted chromosomes at cell division, but the chromosome condensin proteins are also critical.

How is it then that in each cell type only some genes can be set up with all the appropriate transcriptional machinery? Specific transcription factors are clearly important. But RNA is also important. A good example comes from one of the small RNA types. The DNA bases can be methylated, notably cytosine. These methyl marks on cytosine on a stretch of DNA can prevent proteins such as transcription factors binding to DNA. Some small RNAs, small interfering RNAs (siRNAs), can signal the methylation of promoters and prevent gene expression. The role of all noncoding RNA types in controlling the expression of genes in a particular cell type continues to be unraveled.

The methyl marks on DNA are significant in what is known as epigenetics as they can influence gene expression without any sequence changes. The methylation changes are mostly reversible. However, some methylation changes may not be temporary and are inherited. Histone modification, as well as the noncoding RNAs, can also cause epigenetic changes. Epigenetic changes contribute to the formation of different cell types in development due to differences in gene expression. Epigenetics has also been linked to the environment and disease. There is a good biological example of environmental effects in plants, where there is a cold requirement that triggers an epigenetic switch, enabling the induction of flowering in warmer months. Epigenetic changes that occur because of early-life adversity/trauma in children have been linked to subsequent mental illness in adulthood. Clearly these latter findings need to be explored further given the social implications.

Transgenerational inheritance of epigenetic change has been shown in model organisms. Just how important transgenerational epigenetics is in humans continues to be explored given its medical and sociological significance. It is a contested area, given that transgenerational epigenetics is more limited in mammals compared to plants and other animals. A historical study from Överkalix, Northern Sweden, has attracted a lot of attention. The food supply of grandparents in the early twentieth century was linked to mortality rates in subsequent generations. Though DNA methylation has been suggested, it has not been proven. A predisposition to obesity and addictive behavior in mice across multiple generations could not be linked to DNA methylation, but other types of epigenetic change related to histones and noncoding RNAs have been proposed but not yet investigated. The full significance of transgenerational epigenetics remains open. Fundamental research on the inheritance of epigenetic marks continues in mouse models where methylation is induced at specific sites.

Whether regulation by transcription factors and/or noncoding RNAs, a program for a particular cell type must be signaled to set up the cell type. One way of gaining insights into these signaling mechanisms for a cell program is via the study of stem cells. Stem cells are essentially cells that are not specialized for particular functions but with the potential to develop into different types of cells (pluripotent or could be called multipotent cells if more restricted in potential) or all the cells of the organism (totipotent).

A good biological example of pluripotent stem cells is the cells that are at the shoot apex of flowering plants. From these cells, the different cells of the aboveground shoots are produced. In perennial plants, these stem cells will persist throughout the life of the plant, which can be thousands of years in some tree species. A plant zygote derived from the fertilization of the egg cell by sperm cells is totipotent as it is the source of all the cells of the new plant. Plants are rather special in that mature isolated cells in many species when cultured can be reprogrammed to become totipotent stem cells that produce a new plant, a clone. The small molecules that initiate the reprogramming are the plant hormones, particularly auxins and cytokinins. Auxins and cytokinins interact with receptors, which sets in train processes that regulate the transcription of key genes. Cloning plants from mature somatic (nonreproductive) cells in culture was accomplished more than 60 years ago. Cloning animals and mammals is a more difficult proposition. However, John Gurdon in 1966 (with V. Uehlinger) succeeded in using a frog species (*Xenopus laevis*) by building on Gurdon's previous work. Nuclei were transferred from the intestines of feeding tadpoles into a fertilized egg cell, from which its nucleus had been removed, to produce adult frogs. This supported the idea that mature cells still had all the genetic information for the organism and could be reprogrammed to form clones. This influenced the idea that mature human cells could be reprogrammed into pluripotent stem cells. Gurdon shared the Nobel Prize in 2012 with the stem cell pioneer Yamanaka (see later). It was 30 more years before cloning by nuclear transplantation, spearheaded by Ian Walmut, was shown to be possible in mammals. In 1997 with a study in sheep, the nucleus of a mammary gland cell was transferred to an enucleated unfertilized egg cell to produce a cloned sheep, widely known as "Dolly." This is an illustration of the totipotency of nuclei from somatic cells in mammals. In this case, the signals to initiate totipotency in the transferred nucleus come from the egg cell.

In considering stem cells in adult mammals and humans, a good place to start is the stem cells that produce the different types of blood cells in a process known as hematopoiesis. The stem cells for hematopoiesis in adult humans, produced in the bone marrow of adults, are self-renewing pluripotent cells restricted to differentiating into blood cell types. The stem cells develop into two classes, the lymphoid and myeloid cells. The lymphoid cells include B cells, T cells, and NK (natural killer) cells, which are central to the immune system. The myeloid cells ultimately produce several cell types (basophil, eosinophil, erythrocyte, macrophage, monocyte, neutrophil, platelet, dendritic). The erythrocytes or red blood cells are well known, transporting oxygen from the lungs to all body tissues and carbon dioxide back to the lungs. How do these various cell lineages come about? The signals come from the microenvironment in the bone marrow (called the stem cell niche), which is influenced by multiple cell types. The relationship between stem cell renewal and differentiation is accomplished by asymmetric cell divisions where one cell continues the stem cell line and the other daughter cell the new cell type. The signal molecules bind to stem cell receptors and activate a signal transduction pathway, which will program the appropriate genes to form the appropriate cell types. This sequence of events is studied in detail by biochemists and molecular biologists to understand each step of the process.

The understanding of the role of bone marrow in hematopoiesis led to stem cell transplants in the 1950s using bone marrow. Edward D. Thomas and coworkers published this work in 1957 in the *New England Journal of Medicine*, and Thomas received the Nobel Prize for this work in 1990. Bone marrow transplants are now in widespread clinical use. Bone marrow transplants can be carried out using a patient's own bone marrow or bone marrow from a suitable donor. Bone marrow transplants are a way of treating blood diseases such as the blood cancer leukemia.

Stem cells that are pluripotent, with the capacity to be cultured and to form multiple cell types, are valuable for novel investigation strategies and clinical benefits for diseases including type 1 diabetes, Parkinson's disease, and spinal repair in spinal cord injuries. Fundamental research into development and for drug testing would also benefit. Embryonic stem cells have such a pluripotency capacity. These cells (ES cells) were first obtained in the early 1980s from blastocysts (a rapidly dividing ball of cells) of early mice embryos that were several days old. Subsequently in 1998, ES cells were obtained from blastocysts of human embryos by James Thomson and coworkers. Cleavage-stage embryos were produced by *in vitro* fertilization from couples undergoing fertility treatments. These were donated after informed consent and after institutional review board approval. The embryos were then cultured to the blastocyst stage to produce the ES cells.

The production of human ES cells poses ethical problems for human life even if they are produced essentially as a by-product of the clinical treatment of infertility. The ethical concerns accelerated a focus on the use of mature cells, not involving human embryos. This ultimately led to the generation of induced pluripotent stem cells (iPSCs). Shinya Yamanaka (with Kazutoshi Takahashi) in 2006 published in the journal *Cell* the successful generation of iPSCs from mouse fibroblast cells (cells that produce connective tissue). Genes introduced for four transcription factors (OCT3/4, SOX2, C-MYC, and KLF4—the Yamanaka factors) were required for this reprogramming of the fibroblasts into the embryonic state. The same quartet of transcription factors could also reprogram human fibroblasts into iPSCs and was reported in 2007. Human iPSCs have similar properties to ES cells but their availability and lack of potential ethical concerns make them a preferred option.

Given that stem cells can be produced for *in vitro* studies in cell culture, how can they be coaxed into particular cell types? As discussed earlier in this chapter, signals are necessary to set in train specific signaling pathways, which leads to the regulation of the appropriate transcription factors. Some genes need to be switched on while others need to be switched off. Stem cells have been differentiated into a wide range of cell types using a stepwise addition of growth factors and small molecules (e.g., small molecules that influence signaling pathways), suitable culture media, and the appropriate culture conditions. In the process, the stem cells first differentiate into one of the three early embryo germ layers (endoderm, mesoderm, or ectoderm) that give rise to the different organs in the adult. Then, the appropriate cell lineage for a particular cell type is induced. An example of this approach comes from the research on hepatocytes, where the long-term objective is to utilize these cells in the treatment of liver disease without having to use liver transplantation. Human iPSCs produced by Yamanaka factors went through four stages. These were endoderm induction, hepatic cell specification, hepatoblast expansion (proliferation), and maturation of hepatoblasts into hepatocyte-like cells. Another approach to specifying specific cell types is by using gene-specific transcription factors to specify a particular cell type. Again, using hepatocytes as an example, stem cells are differentiated into endoderm, then specific transcription factors used to produce the hepatocytes. It has been possible to produce a wide range of cell types from cultured stem cells. Examples in addition to hepatocytes include neurons, pancreatic cells, hematopoietic cells, and cardiac cells.

While the utilization of cultured cell types has yet to reach its full potential, there are many applications already in use. Patient-derived iPSCs can be used for disease modeling and developing therapeutic strategies. One example is the use of motor neurons from iPSCs derived from the fibroblasts of patients with spinal muscular atrophy (SMA). Neurons-derived iPSCs have also been used to model Alzheimer's disease and Parkinson's disease. These patient-derived iPSC models allow disease pathologies in a culture dish, facilitating drug discovery and screening of existing drugs.

The potential use of stem cell technology in regenerative medicine has created wide interest and excitement. One active area that has involved clinical trials with ESCs and iPSCs is in the production of RPE cells (retinal pigment epithelium cells) for the treatment of macular degeneration. Macular degeneration is an eye disease, which can lead to blindness, where there is a progressive deterioration of light-sensing photoreceptors of the retina. Transplanted sheets of RPE cells have stopped macular degeneration and led to improved vision. The technique, which includes clinical trials, is making steady progress. Other examples being investigated are the use of cardiomyocytes, produced from pluripotent stem cells, in the treatment of heart disease and stem cells in the treatment of spinal cord injury. One of the advantages of iPSCs using the patient's own cells is that there is no need for long-term immune suppression to preserve the transplanted cells. However, despite effective reprogramming technologies there are potential problems with the integration of exogenous genetic factors, including the use of oncogenes. Research continues on eliminating these potential problems by using strategies that eliminate or reduce the use of approaches that utilize DNA integration into the genome. In this regard, there is progress in using small molecules and growth factors in reprogramming. There is still a lot more to learn about gene regulation in determining cell types and disease development, but iPSCs represent a paradigm shift in providing new investigative strategies.

In considering pluripotent stem cells in the context of biology and potential clinical therapies, the end products are cell or tissue types, in two dimensions. Organs are built up of many cell types and are 3D. There has been recent progress in producing 3D organoids. Organoids are miniaturized versions of adult organs, with many of the properties of the organ produced by normal development. Pluripotent stem cells have been used to generate organoids, including brain, kidney, liver, and lung. These structures can be produced in culture by controlling the media, the way the cells are cultured and utilizing the self-organizing capacity of the different tissues. The 3D organoids provide improved models for disease and drug testing in the relevant spatiotemporal context.

What does the future hold for stem cell biology and clinical use? There will almost certainly be progress in the understanding of the molecular biology of how genes are regulated to produce different cell types. This increased knowledge will facilitate new approaches to disease treatment. Disease modeling will be valuable. An interesting example is the modeling of Zika virus infection using human brain organoids. The clinical use of stem cell therapy holds great promise but will need well-developed animal models to inform research approaches on quality and reproducibility.

Stem cell research is advancing at a rapid pace and new understanding and applications of stem cells are being generated. "Synthetic" embryos, or more biologically known as stem cell-based embryo models, without using eggs or sperm, have been grown from mouse stem cells. More recently, human stem cells have been grown to a stage equivalent to embryos 14 days after fertilization. These embryo models, however, skip the blastocyst and early implantation stages. This latter work is illustrative of the progress and potential of stem cell research, and the ethical concerns involved. How far embryo models should be developed in search for an understanding of embryo development remains an important question for society.

Further Reading

Clevers H. Modeling development and disease with organoids. Cell. 2016;165:1586–97.

Fitz-James MH, Cavalli G. Molecular mechanisms of transgenerational epigenetic inheritance. Nat Rev Genet. 2022;23:1–17.

Gunsilius E, Gastl G, Petzer AL. Hematopoietic stem cells. Biomed Pharmacother. 2001;55:186–94.

Gurdon JB, Colman A. The future of cloning. Nature. 1999;402:743–6.

Hong JC. General aspects of plant transcription factor families. In: Gonzalez DH, editor. Plant transcription factors. Boston: Academic Press; 2016. p. 35–56.

Kornberg RD, Lorch Y. Primary role of the nucleosome. Mol Cell. 2020;79:371–5.

Myers LC, Kornberg RD. Mediator of transcriptional regulation. Annu Rev. Biochem. 2000;69:729–49.

Shi Y, Inoue H, Wu JC, Yamanaka S. Induced pluripotent stem cell technology: a decade of progress. Nat Rev Drug Discov. 2017;16:115–30.

Workman JL. Nucleosome displacement in transcription. Genes Dev. 2006;20:2009–17.

Zakrzewski W, Dobrzyński M, Szymonowicz M, Rybak Z. Stem cells: past, present, and future. Stem Cell Res Ther. 2019;10:1–22.

6

The Cell Division Cycle and the Biology of Cancer

As part of the cell division cycle, DNA is duplicated and then one copy is distributed to each new cell. The cell division cycle is divided into phases. G1, S, and G2 occur in interphase prior to chromosome separation in mitosis (M). S phase is the DNA duplication phase. The initiation of DNA replication, replication forks, and Okazaki fragments is discussed. Progression through the cell cycle is dependent on cyclin-dependent kinases (CDKs) and cyclins. Cancer occurs when there is uncontrolled cell division. The "Guardian of the Genome" the p53 protein tumor suppressor is a critical regulator, and its mutation occurs in more than 50% of cancers. The mutation or epigenetic change of oncogenes can also lead to cancer. Stem cells are good targets for cancer, but mature cells can dedifferentiate and also be a source of cancer cells. The cancer stem cell idea is discussed, particularly in the case of cancer recurring after remission. Drugs developed for chemotherapy increasingly strive for selectivity to inhibit key proteins involved in signaling the promotion of cell division and cell cycle regulation.

In this chapter, the focus is on the eukaryotic cell (cells like ours). The growth of all eukaryotes is dependent on cell division. All our component organs grow by cell division. As discussed in Chap. 5, the characteristic cells of an organ are determined by the suite of expressed genes. In the cell division cycle, a parent cell divides into two and then each

R. J. Rose, *DNA Know Thyself*,
https://doi.org/10.1007/978-981-97-8381-6_6

daughter cell grows to the size of the parent cell. As part of the cell division cycle, DNA is duplicated and then one copy is distributed to each new cell. The DNA is in the nucleus that is bound by a double membrane. The DNA of the human cell is organized into 46 chromosomes, which are maintained in each cell. Each chromosome, in the unfolded state, is a long DNA molecule wrapped around the nucleosomes like beads on a string (Chaps. 4 and 5). When mitosis is reached, the chromosomes are precisely folded to form rod-like structures.

The cell division cycle is organized into discrete phases. The central part of the cell cycle is the DNA replication phase where each chromosome is duplicated. The DNA duplication phase is known as the S phase. Prior to the initiation of the S phase is the G1 (Gap1) phase and after the S phase is the G2 (Gap2) phase. The G1, S, and G2 phases are collectively known as interphase. The end of the G2 phase occurs when the chromosome separation process begins, signaling the beginning of mitosis (M), known as prophase. During mitosis, the duplicated chromosomes separate into two separate chromosome sets at each pole of the cell. The cell then divides into two between the two chromosome sets in a process known as cytokinesis. At the end of cytokinesis, each daughter cell will then have a complete set of chromosomes.

How does the cell cycle progress through each phase in a regularly timed process? The understanding of the general mechanism of cell cycle progression is now reasonably well understood. An important contribution to increasing this knowledge came from using the yeast species *Saccharomyces cerevisiae*, a single-celled eukaryote. *S. cerevisiae* has a special feature in the way it undergoes cell division. This yeast divides by budding, with the bud continually increasing in size until the size of the bud equals the size of the mother cell. Cell division then occurs. The size of the bud provides a visual marker of the position in the cell cycle. Leland Hartwell used these cells to produce mutants to discover genes required for cell cycle progression. Another researcher, Paul Nurse, in a similar vein, used another yeast *Schizosaccharomyces pombe,* known as the fission yeast. This yeast grows at a constant rate until it doubles in size, different lengths correspond to different stages of the cell cycle. When cell size is doubled, the cell divides into two. The Nurse group obtained mutants that divided when the cells were smaller than normal (known as "*wee*" mutants, a Scottish term for small). Yeast mutants were critical in elucidating cell cycle mechanisms.

The Leland Hartwell, Paul Nurse, and Tim Hunt groups provided a basis for the control of the cell cycle. The three investigators received the Nobel Prize for their discovery of the key regulators of cell division cycle progression. Two types of proteins are involved, the cyclin-dependent kinases (CDKs) and the cyclins. The CDKs were discovered in yeast and were necessary for G1-S and G2-M transitions. The first cyclins were discovered by Tim Hunt using fertilized sea urchin eggs. The CDKs are activated by cyclins. CDK-cyclin complexes then phosphorylate target proteins by adding a phosphate group ([P]). The targeted proteins set in train mechanisms that drive a particular phase of the cell cycle. Different cyclins activate different phases of the cell cycle; G1-S, S progression and G2-M by interacting with a CDK. In yeast, a single CDK is activated by multiple cyclins, with specific cyclins responsible for initiating specific cell cycle events. The cyclins form and degrade in each cell cycle. Yeast were magnificent models for developing an understanding of the cell cycle. The replication of DNA and the partitioning of the chromosomes are characteristic of eukaryotic cells, so the yeast findings are applicable to plants, animals, and humans. The engines of the cell cycle in eukaryotes are then G1 cyclin-CDKs, S-phase cyclin-CDKs, and mitotic cyclin-CDKs acting sequentially. In mammalian cells including humans, multiple CDKs and cyclins are involved, with specific combinations determining each cell cycle transition point.

DNA replication is a central part of the cell cycle where each chromosome is duplicated. It is the elegant DNA structure that facilitates this. Each of the DNA strands of the double helix serves as a template to form a new DNA double helix, following the base-pairing rules. How do the G1-cyclin-CDKs initiate the DNA replication process? In G1, there is a point known as a checkpoint or restriction point that needs to be traversed if the S phase and the cell cycle are to proceed. Prior to this point, a protein, known as the RB (retinoblastoma) protein, blocks any further progress. When the CDKs phosphorylate ([P] addition) this RB protein, the block is released. The phosphorylated RB protein activates the E2F transcription factors (Chap. 5 discusses transcription factors). While this is a key point in the initiation of S, there are many connected proteins involved in this restriction point release. The *RB* gene (the retinoblastoma gene) is known as an oncogene, as a mutation can cause cancer, and this will be discussed later.

When E2F transcription factors are activated, what is involved in setting up DNA replication? One important point to note is that replicating the chromosome does not simply involve starting (called an origin) at one end and working to the other end. The chromosome is so long that there are multiple origins, probably to speed up the process, where replication initiates and then all the replicated portions join. It is somewhat surprising that while each origin usually contains A-T-rich sequences they do not have a common sequence and possibly there is something about a chromosome with its associated nucleosomes that may need to be factored in. To understand the DNA replication initiation process, we need to consider a single origin. At the origin, the two strands are unwound to separate and form a "bubble"-type structure with a fork at either end, which can be visualized in the electron microscope. The unwinding of the DNA is carried out by enzymes called helicases that give access to the single DNA strands, which allows for base-pairing of individual nucleotides and their polymerization, forming new double-stranded DNA. Understanding DNA replication centers on what happens at the replication forks. As the fork moves forward, there is a leading strand and a lagging strand necessitated because of the antiparallel nature of the DNA double helix. In the case of the lagging strand, the DNA is replicated in small pieces called Okazaki fragments, which are joined together to form a continuum. Okazaki fragments are named after the scientists Reji Okazaki and Tuneko Okazaki, who proposed their role in DNA replication in 1968. The DNA polymerases are enzymes that carry out DNA replication in leading and lagging strands. Current evidence supports a model where DNA polymerase ε carries out replication of the leading strand and different DNA polymerases (DNA polymerases α and δ) are involved in the discontinuous replication involving Okazaki fragments in the lagging strand. Cellular DNA replication mechanisms are highly conserved in all organisms, consistent with the original Watson and Crick model of the DNA molecule. The replication of the antiparallel double-stranded DNA involves continuous synthesis of the leading strand and a discontinuous synthesis of a lagging strand involving Okazaki fragments.

The completion of the replication of the chromosomes by the end of the S phase signals the next major step in the cell cycle, which is the separation of the duplicated chromosomes by mitosis. The duplicated chromosomes start to condense (fold) and in late G2 become visible in the light microscope at the beginning of prophase. Appropriate cyclin-CDKs need to be activated to proceed into mitosis. An important checkpoint is the end of DNA replication; if this is not completed or there is DNA damage, then the specific cyclin CDKs are not activated and mitosis will not proceed. DNA replication must be completed, and any DNA damage repaired before proteins for the mitotic process are activated and mitosis begins. The transcription factor gene p53 is critical in orchestrating DNA repair. As a transcription factor, the p53 protein can activate a plethora of genes to facilitate its function. In response to DNA damage, the cell cycle is slowed, and the DNA is repaired using DNA repair enzymes.

How are the condensed duplicated chromosomes separated in mitosis? As the cells move from prophase to metaphase, the nuclear envelope is lost and the spindle microtubules from both poles attach to the kinetochore of each sister chromosome. In animal cells, the spindle emanates from the centrosome at each pole. The centrosome is duplicated in interphase. In plant cells, there are no centrosomes, but there are microtubule-organizing systems at the poles. The spindle must be attached to the kinetochore of all chromosomes before mitosis will proceed to anaphase. The kinetochore is the protein complex at the centromere where sister chromosomes are tightly connected. Key to the next step is the anaphase-promoting complex/cyclosome (APC/C) that triggers processes required for separating the chromosomes. Each set of chromosomes moves to the opposite spindle pole by shortening of the microtubules at the kinetochore and spindle poles.

Telophase follows anaphase, with the chromosomes destined for the two daughter cells decondensing and the nuclear membrane reforming. The next step is cytokinesis, where the cell is divided into two. In vertebrates, including humans, the cell constricts into two, in plants a new cell wall forms between the two daughter cells. Each daughter cell will enter G1, and the cell cycle commences again. A cell may exit the cycle to differentiate into a particular cell type. In this case, the RB protein is not activated and cells do not progress into the S phase.

Mitochondria also contain DNA (Chaps. 2 and 3) and do not arise de novo, so each daughter cell must contain mitochondria. Mitochondria are dynamic organelles, undergoing fusion and fission in the cell cycle. The net effect is to result in increased numbers of mitochondria distributed throughout the cell with the help of the cytoskeleton (microfilaments and/or microtubules depending on the eukaryote) so when the cell divides each daughter cell will receive similar numbers of mitochondria. Plant cells also contain the DNA-containing chloroplasts as well as mitochondria. Chloroplasts need to divide in the cell cycle (they do not undergo fusion) and then distribute to the new cells.

How is the cell cycle connected to cancer? Cancer is the disease where cells undergo uncontrolled cell division. In some cases, these proliferating cells may spread to other parts of the body in a process known as metastasis. In understanding cancer and potential treatment, a knowledge of the cell cycle is essential. In the overview presented here, the main aspects of the cell cycle and its regulation are presented. While the main genes and their proteins have been featured, many genes feed into the major controllers. A good place to start in understanding cancer is with the p53 protein, which is widely known as the "Guardian of the Genome" and as a tumor suppressor. As previously indicated, p53 acts at the checkpoint at the end of the S phase.

DNA replication errors can be corrected by p53 activating DNA repair genes at this stage. If DNA damage is extensive, p53 can activate genes that lead to apoptosis, the process of programmed cell death. This process is important in preventing cancer, particularly where there are multiple mutations. If p53 mutations lead to a defective p53, then cancer is a likely outcome. More than half of human cancers harbor p53 mutations. This has led to research that aims to correct p53 using small molecules as well as other approaches to reactivate p53 activity. This research is ongoing, with some drugs at the clinical testing stage.

The RB protein as discussed above is a universal cell cycle regulator with a key role in signaling the initiation of DNA replication to start the cell cycle. *RB* is an oncogene, and if its function is disrupted, there is unregulated cell cycle progression and tumor growth is promoted. Mutations in *RB* and *E2F* family members can impact the initiation, progression, and prognosis of cancer. Impaired p53 genes, or genes that impact their function, illustrate the significance of the cell cycle in a full understanding of cancer. Another group of genes that are important are those responsible for DNA repair. It is one thing to signal that DNA repair is needed, it is another to initiate repair. DNA repair genes are sometimes called caretaker genes. Hereditary deficiencies in the well-known breast cancer genes *BRAC-1* and *BRAC-2* can cause cancer. These tumor suppressor genes are DNA repair genes, particularly for double-strand DNA breaks.

Another type of oncogene is *KRAS*, which acts as an on/off switch in signaling pathways, including signaling involved in the promotion of cell growth. If a *KRAS* mutation prevents the cell growth switch from turning off, then cell proliferation can continue uncontrollably. *KRAS* is the most mutated oncogene in human cancer, making it a high-priority therapeutic target. *KRAS* mutations are commonly involved in pancreatic, lung, and colorectal cancers.

Cancer is a particularly complex disease because many cell cycle and DNA repair genes (as well as genes that signal cell proliferation) can be causal and there is metastasis due to cells breaking away from the tumor. Tumors in different organs have their own characteristics, whether primary or secondary due to metastasis. Primary cancers usually initiate in cells that are actively dividing so that the mutation(s) can be passed on to many progeny cells. How does this happen in a mature individual? Stem cells differentiate into particular cell types, but a population of stem cells require continuous renewal by cell division. Examples of adult stem cells are in the bone marrow where hematopoiesis occurs to continually replenish blood cells (see Chap. 5), cells that form skin, and cells that line the small intestine. Acute myeloid leukemia is an example of a cancer caused by the myeloid line of hematopoietic progenitor cells that normally produce blood cells including red and white blood cells, and platelets. The massive proliferation of developing stem cells disrupts normal blood cell production. Skin stem cells are self-renewing and give rise to the different cell lineages that form mature epidermal cells. These stem cells and their lineages prior to differentiation are prone to cancer, particularly in response to the UV light in sunlight causing DNA mutations. Nonmelanoma skin cancers, including basal cell (BCC) and squamous cell carcinomas (SCC), are the most common human cancers. SCCs originate from the stem/progenitor cells of the basal layer of the epidermis, while BCCs may be similarly derived as well as from hair follicle stem cells. Melanoma cancers are much more lethal than BCCs or SCCs and derive from dedifferentiated melanocytes or stem/progenitor melanocyte cells. The lethality of melanomas is due to their ability to metastasize. Stem cells in the intestinal crypts of adults fuel the constant replenishment of the intestinal epithelium. The intestinal stem cells are the most potent in forming colorectal cancer, but more differentiated progenitor cells may also serve as the origin of cancer cells.

Stem cells have their special niches with a characteristic microenvironment that is influential in their behavior, as well as intracellular factors influencing DNA integrity. The stem cells in adults considered above in the context of cancer are involved in replacing cells with a rapid turnover requiring continuing replacement, as in blood, skin epidermis, and the intestinal epithelium. However, in mammals, including humans, adult stem cells have been identified in virtually all tissues to facilitate cell replacement in response to injury or disease. For example, primary liver cancers can be derived from transformed liver stem cells, progenitor cells, and dedifferentiated cells derived from cells such as hepatocytes. Less powerful stem cell niches occur in the heart and nervous tissue and are unable to repair severe damage.

An understanding of cancer cell biology requires a detailed knowledge of the cell cycle and key controlling genes as well as stem cell biology. In cancer initiation, adult stem cells can be cells of origin as well as dedifferentiated cells that can take on the function of induced pluripotent stem cells. Cells that are progenitors of tumors can be thought of as being derived from cancer stem cells (CSCs). A CSC can be derived from preexisting stem cells or stem cells derived from dedifferentiated mature cells (like induced pluripotent stem cells of Chap. 5). These CSCs may exist in small numbers and a given CSC can form a tumor in response to appropriate mutations. The existence of a group of CSCs helps explain how a tumor can be under remission, then another CSC may be activated to form a new tumor.

Cancers are ultimately due to negative changes in gene expression. This is commonly due to mutation, but epigenetic modification of oncogenes and tumor suppressor genes can also contribute. Epigenetic change does not involve DNA sequence change (see Chap. 5), opening new possibilities for the development of new types of drugs directed against DNA methylation and histone modifications. Drugs directed against DNA methyltransferases have been approved by the US Food and Drug Administration. Similarly, there are drugs directed against histone modifications such as histone deacetylase. As understanding of DNA mutation and epigenetic change leading to cancer initiation, progression, and metastasis in different tissues increases, there is an increasing library of cancer drugs that can be used in chemotherapy.

Increased knowledge of the cell division cycle and its regulation provides the basis for targeted chemotherapy. Earliest strategies were DNA-centered, damaging DNA and its replication leading to apoptosis. Examples are the DNA synthesis inhibitor 5-fluorouracil (a uracil analog) and mitomycin C (an antibiotic), which damages DNA. Inhibitors of CDKs is another strategy to stop cell division. The general idea is to stop the division of the rapidly proliferating cancer cells. However, these strategies also damage nondividing normal cells. This is in principle the same approach as radiation that causes double-stranded DNA breaks followed by apoptosis. Improved radiation technology to precisely target the cancer cells has gone hand in hand with improved approaches to chemotherapy. Current cancer drug development particularly focuses on inhibiting defined targets in signaling pathways promoting cell division to selectively kill cancer cells. Various kinases involved in signaling are common targets. These drugs are usually small molecules that block the activity of the protein involved. In the past 20 years, many small molecules have been used or are being developed, leading to an increased library of cancer drugs that can be directed against cancers of specific organs.

Metastasis is less well-understood compared to cell proliferation and is more likely to be fatal. Only a small subset of cells in primary tumors can survive in the circulatory system and distant organs, and then continue tumor growth. Tumors that develop from CSCs do not lead to a tumor of a single cell type; it is heterogeneous with a cellular ecosystem. For a tumor to grow rapidly, nutrients have to be provided via blood vessels that permeate the tumor. The heterogeneity of cellular metabolism in tumors provides the potential for metastasis. It is now possible to obtain the different cell types and analyze their gene expression changes, increasing the details of tumors in different parts of the body. What this is leading to is a much more detailed understanding of what molecules to target in chemotherapy. There is cause for optimism in that refined treatments will be developed based on the detailed knowledge of tumor initiation and how tumors develop. In the consideration of the cell cycle and cancer biology in this chapter, immunological strategies for cancer therapy have not been considered, which is also being pursued using DNA-related technology (see Chap. 13).

Further Reading

Afify SM, Seno M. Conversion of stem cells to cancer stem cells: undercurrent of cancer initiation. Cancers. 2019;11:345.

Baksh SC, Finley LW. Metabolic diversity drives cancer cell invasion. Nature. 2022;627-8:627.

Brady CA, Attardi LD. p53 at a glance. J Cell Sci. 2010;123:2527–32.

Dick FA, Rubin SM. Molecular mechanisms underlying RB protein function. Nat Rev. Mol Cell Biol. 2013;14:297–306.

Hartwell LH, Culotti J, Pringle JR, Reid BJ. Genetic control of the cell division cycle in yeast: a model to account for the order of cell cycle events is deduced from the phenotypes of yeast mutants. Science. 1974;183:46–51.

Li Y, Jin J, Bai F. Cancer biology deciphered by single-cell transcriptomic sequencing. Protein Cell. 2022;13:169–79.

Mannino G, Russo C, Maugeri G, Musumeci G, Vicario N, et al. Adult stem cell niches for tissue homeostasis. J Cell Physiol. 2022;237:239–57.

Nurse P. A long twentieth century of the cell cycle and beyond. Cell. 2000;100:71–8.

Okazaki T. Days weaving the lagging strand synthesis of DNA—A personal recollection of the discovery of Okazaki fragments and studies on discontinuous replication mechanism. Proc Jpn Acad Seri B. 2017;93:322–38.

Yamada Y, Haga H, Yamada Y. Concise review: dedifferentiation meets cancer development: proof of concept for epigenetic cancer. Stem Cells Transl Med. 2014;3:1182–7.

7

Generation Upon Generation

The title of this chapter is the same title as Chap. 12 of Jacob Bronowski's book The Ascent of Man, *which accompanied the 1973 television documentary series. He points out that "sex produces diversity, and diversity is the propeller of evolution." While there is DNA exchange in bacteria in what is called parasexuality, sexual reproduction evolved in the eukaryotes. The meiosis process allows for the production of the haploid gametes, sperm, and egg. Specific transcription factors signal entry into meiosis rather than mitosis. Crossing over, recombination and reduction division occurs in meiosis I. At the end of meiosis I, each cell contains a single set of chromosomes consisting of two chromatids. There is an independent assortment of the maternal and paternal chromosomes. At meiosis II, the chromatids separate so that each new cell becomes a haploid gamete. The male and female gamete cells come together in the process of fertilization to form the zygote and restore the diploid chromosome number and produce the next generation. The essential immortality of gamete DNA as opposed to somatic DNA is discussed, as well as trends in human fertility.*

The first sexual eukaryotes were single-celled protists about 1.5 billion years ago and are associated with the evolution of the diversity of eukaryote life that exists on this planet. In the last 50 years, a more sophisticated understanding of evolution and sexual reproduction has come

R. J. Rose, *DNA Know Thyself*,
https://doi.org/10.1007/978-981-97-8381-6_7

with the development of genomics and quantitative genetics. However, mutation, recombination, and independent assortment are responsible for the variation that provides the raw material for natural selection. Recombination and independent assortment occur in the formation of the gametes, sperm, and egg in the process of meiosis. Meiosis is the process where the number of chromosomes is halved in the formation of the gametes (sperm and eggs) and where crossing over and subsequent recombination, and independent assortment of homologous chromosomes occur. This chapter focuses on the current understanding of the processes that DNA undergoes in meiosis, the formation of gametes and fertilization. There is also the question of how DNA integrity has been maintained generation upon generation.

Meiosis occurs in specialized cells whether in plants or animals. Early inheritance studies by Gregor Mendel were in plants, published in 1866. In plants, the reproductive cells are part of the flower. Meiosis and the production of pollen containing the sperm cells occur in the anthers. Meiosis and the production of eggs occur in the ovary, at the base of the carpel. In humans, meiosis and sperm production occur in the testes, while meiosis and egg production occur in the ovary. Meiosis is a type of cell division where there are two cell divisions and one round of DNA replication with four gamete cells as the final product. Each gamete contains a single chromosome set, halving the number in the parent cell. Fertilization of the egg and sperm restores the chromosome number in the offspring. In considering meiosis, it is convenient to divide it into phases, meiosis I and meiosis II.

The first division of meiosis, meiosis I, is an impressive process. In meiosis I just as it is in mitosis, there is a preceding round of DNA replication. However, if a cell is to enter meiosis and not mitosis, this is signaled prior to DNA replication. It is a commitment to the formation of the reproductive gametes, a key event in the life cycle. While the physical mechanism of meiosis is similar across species, the transitioning to meiosis is not well conserved, and especially in mammals there are differences between the sexes with hormonal difference coming into play. There are no human cell lines to enable basic research. However, it is useful for research purposes that meiotic entry and progression are similar in mice and humans. In plants, understanding meiosis is also of great importance for crop breeding and determinants of crop yield, but again much more research is required. A recent strategy that has been used in maize is single-cell transcriptomics, which is beginning to define the gene networks initiating and progressing meiosis. In considering the initiation of meiosis, there are two main factors. The first is the initiation of DNA replication, and the second is the determination that this is meiosis with its special characteristics. In relation to the initiation of DNA replication, it is like that involved in mitosis with some changes in the specific genes involved. The signaling that this is a meiotic event is analogous to cell fate determination from stem cells with the appropriate transcription factors activating the appropriate set of genes. In male and female mammalian germ cells, two transcription factors play critical roles in the mitosis-to-meiosis cell fate transition. These transcription factors are STRA8 and MEIOSIN, which are expressed in response to retinoic acid signaling.

When the chromosomes start to condense after DNA replication and when prophase I starts, subsequent events are peculiar to meiosis I. At this stage, each chromosome has duplicated, and it is easier to refer to the replicated chromosome as consisting of two chromatids. Homologous chromosomes (the same chromosome from each parent) then pair along their length in a process called synapsing.

How do the chromosomes find each other? The tethering of telomeres (repeated sequences at chromosome ends, also see Chap. 8) to the inner nuclear envelope has a central role. Homologous chromosomes occur in similar locations or territories, and it is likely that they are close by with both tethered to the inner nuclear envelope via telomeres, with LINC complexes connecting the telomeres to the cytoskeleton near the outer nuclear membrane. The cytoskeleton can drive the mobility of the chromosomes so that alignment of homologs is facilitated, joined by the synaptonemal complex in the synapsing process. As this is occurring, crossing over between nonsister chromatids occurs. What this enables is an exchange of portions of the nonsister chromatids. This can be explained by referring to our own situation. We have inherited one set of chromosomes from our mother and one from our father. At prophase I of meiosis, DNA has replicated so that for homologous chromosomes there are two maternal chromatids and two paternal chromatids. As a result of crossing over between a maternal and paternal chromatid, two of the chromatids are now a mixture of maternal and paternal chromatids and are called recombinants. See the diagram below (Fig. 7.1).

How does this crossing over occur? At the DNA level, there is the formation of a double-strand break in both chromatids followed by DNA repair at the crossover point. These precise DNA exchanges are helped by the synaptonemal complex aligning the homologous chromosomes. There is at least one crossover and recombination per homologous pair of chromatids but there may be more than one.

With the crossover underway, the chromosomes move to metaphase I at the equator of the cell. The nuclear membrane has disappeared at this stage. It is at this point that there is a major difference to mitosis. The chromatids of the replicated chromosome do not separate but go to the same

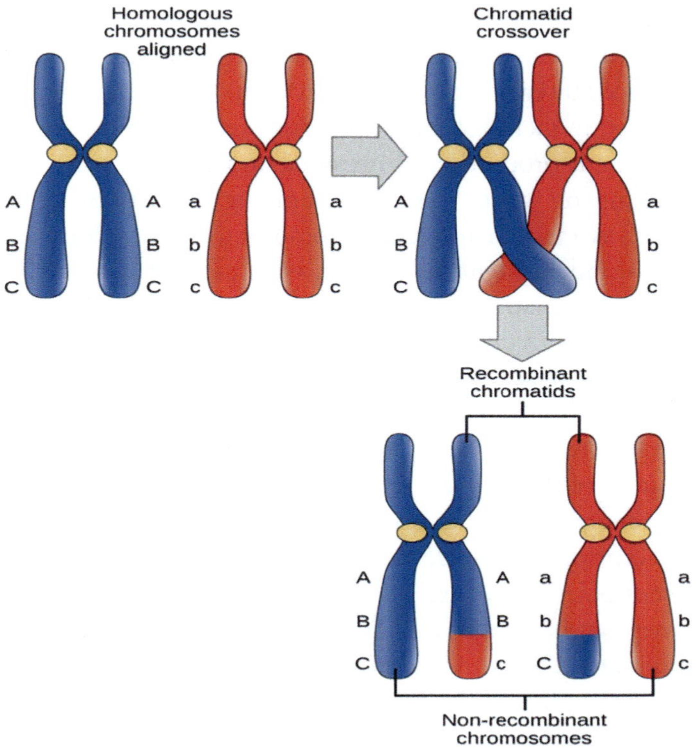

Fig. 7.1 Crossing over between homologous chromosomes. Image source: By CNX OpenStax CCBY4.0, via Wikimedia Commons. See https://commons.wikimedia.org/w/index.php?curid=49926315 for image and attribution

pole, causing a halving of the chromosome number per cell. Microtubules are involved in pulling the two chromatids to the pole in anaphase I. The centromeres (cohesion site between sister chromatids) do not divide, and the microtubules are attached to the kinetochores (a protein complex located at the centromere), which are oriented to ensure that both chromatids go to the same pole, and in the process the crossing over is completed. Telophase and cytokinesis follow so that two daughter cells are formed with half the chromosome number in each cell. Each chromosome still has two chromatids. In addition to the variation that is introduced by crossing over, the way each chromosome is partitioned in meiosis I means that the final gametes will have a different proportion of maternal and paternal genes depending on which pole the homolog is partitioned to, which is random for each chromosome. This is Mendel's law of independent assortment, which introduces another source of variation.

Once the reduction division is completed in meiosis I, meiosis II proceeds with the cells moving rapidly through prophase II into metaphase II in each cell. At metaphase II, the chromosome with two chromatids is aligned at the equator of the cell. The microtubules attach to the two kinetochores, this time oriented in different directions and the centromeres divide as the chromatids are pulled to different poles of the cell in anaphase II. Telophase and cytokinesis ensue in each cell. This means starting from one cell four gamete cells are produced that will form the sperm or egg depending on the parent.

The gamete cells come together in the process of fertilization to form the zygote and restore the chromosome number and produce the next generation. The plant and animal kingdom do this in different but conceptually similar ways. In plants, the sperm cells are packaged in pollen. The pollen is transmitted to the stigma of the carpel in a myriad of different ways depending on the species, whether self-pollinated involving one plant or cross-pollinated involving two plants. The pollen after landing on the stigma germinates to produce the pollen tube, which carries the sperm cells through the style to the ovule in the ovary. One sperm fertilizes the egg housed in the embryo sac of the ovule; the other sperm fertilizes the central cell in the embryo sac. Egg fertilization results in the formation of the zygote while the central cell fertilization produces the endosperm, which provides nutrition for the embryo. Double fertilization is unique to plants.

The biology of fertilization in humans also involves the delivery of the sperm to the egg cell for fertilization. Human sperm, delivered at copulation, is designed for swimming in the reproductive tract of the female to meet with the egg (oocyte) in the fallopian tube. The sperm is quite a complex cell formed by differentiation of the male gamete in spermiogenesis. The sperm head contains the nucleus and the acrosome that contains enzymes involved in fertilization with the egg, and a flagellum tail that propels the sperm. Mitochondria are present in the mid-piece. The chromatin in the nucleus is highly condensed. In humans, there are two ovaries, one on either side of the uterus. A reproductively mature woman releases a single egg each month, which passes into the fallopian tube. The fertilized egg cell in plants and in humans contains mitochondria, which contain a small number of critical genes. In humans, the mitochondria come from the egg cell; they are maternally inherited. In human sperm, mitochondria are present to provide energy but do not survive fertilization. In plants, both chloroplasts and mitochondria are mostly maternally inherited. Why the DNA-containing organelles are maternally inherited is unclear; it is an active process to prevent paternal inheritance. One possibility is that the male gamete on the pathway to fertilization faces environments that could contribute stress, which may have negative effects on mitochondrial (and chloroplast) DNA. In any cell, there are multiple organelle DNA copies and a cell biology that facilitates quality and variation, but mechanistically quite different from the nuclear genome.

The zygote with the combined parental genomes and the mitochondria from the female houses all the genetic information to produce the offspring. In humans, the sex of the offspring is determined by the sex chromosomes, an X and a Y will produce a male and an XX a female. This means each offspring has 2 sex chromosomes and 44 chromosomes termed autosomes. It is interesting to follow the inheritance of the autosomes through two generations. We inherit 50% of our genes (or 22 autosomes) from each parent, with each chromosome of a pair derived from one parent. If we consider our grandparents, we might expect to inherit 25% of the genes from each grandparent, with each grandparent contributing 11 autosomes. But, due to independent assortment, each grandparent will contribute variable number of chromosomes from the grandmothers or grandfathers. This means that rather than a grandparent contributing 25% of the autosomal genes could (but definitely unlikely) contribute up to 50% due to independent assortment in meiosis I. This consideration does not consider crossing over making some chromosomes "hybrids" of two grandparents. Overall, this means that each of our four grandparents mostly contribute different amounts of DNA.

Generation after generation gametes and zygotes are produced. Through the sexual reproductive process, DNA is essentially immortal. The somatic cells of the organism ultimately lead to aging and death but the transmitted DNA via the gametes goes on to form the next generation. In the case of humans, this has been going on for about 200,000 years. In somatic cells, DNA damage has a central role in the aging process. The reproductive DNA also undergoes positive and negative mutations but continues generation after generation. DNA repair occurs in both somatic and germ cells. Somatic cells often respond by arresting the cell cycle to allow time to carry out DNA repair. While germ cells undergo similar repair processes, different outcomes are favored. In germ cells, error-free repair mechanisms and apoptosis are favored. Apoptosis completely removes cells with damaged DNA. In human germ cells, there are characteristics of gamete formation that contribute to protective mechanisms for the inheritance of quality DNA. The oocyte is protected by surrounding cells and the ovarian environment in the maturation process. The oocyte packaging of DNA is histone-based as in somatic cells. In mature sperm, the DNA is packaged differently with very highly condensed DNA being achieved using protamines, which replace the histones. The highly condensed nucleus in the head of the sperm makes it more hydrodynamic and adds protection to any stress due to environmental insults. Essentially DNA damage response mechanisms are optimized for the development of spermatozoon, oocyte, and zygote. Another aspect of human gamete biology assisting genome integrity is the quality of the gametes. In males, it has been estimated that there are more than 40 million sperm per ejaculate and 525 billion produced in a lifetime, so damaged sperm would unlikely proceed to fertilization. Though oocyte production is different with one oocyte released at ovulation, there are 1–two million oocytes at birth and 300,000 at puberty.

Considering DNA integrity in germ cells and somatic cells more generally, it can be argued that the sexual reproduction process involving the production of gametes and subsequent fertilization is a very energy-intensive process. This is an advantage for the species but comes at the expense of the development and death of the somatic cells of the individual. Experiments have been carried out using the zebrafish to test this theory. The zebrafish has become a model vertebrate organism. Among its experimental qualities, it has been sequenced, it has a rapid generation time, and as fertilization is external all developmental stages are accessible. The "expensive germline hypothesis" was tested by comparing the presence and absence of the germline on somatic repair.

Germline-free male fish were produced by knocking down a gene using an antisense approach preventing germ cell formation but otherwise maintaining the male phenotype. Male fish were exposed to sublethal low doses of ionizing radiation, which affected both the soma and germline. What was then tested was the capacity of the fin to regenerate after partial removal. The recovery of the somatic cells of the fin occurred substantially faster in the germline-free fish. Under irradiation, germline-carrying fish invested much of their resources into germline repair. Germline-free fish, on the other hand, have no such costs and restore soma cells at a faster rate.

Overall, there are very good protective mechanisms that essentially make DNA transmitted through the reproductive process immortal. However, there are concerns about current trends in human reproduction. Though debated, there is a well-documented decline in human fertility (i.e., biological, not economic or behavioral) in Europe starting from around the 1900s. In countries where industrialization and economic growth started later, fertility decline started within the last five decades. Sperm counts have halved in the past 50 years in the industrialized west and east, together with a decline in sperm quality and an increase in testicular cancer. In today's society, there is decreasing selection for high-fertility genotypes, and this is being increasingly exacerbated by assisted reproduction technologies. While lifestyles may be a contributing factor, environmental factors appear to be particularly important. It has been suggested in a comprehensive 2022 review by Skakkebæk and coauthors that declines in fertility rates might be linked to chemicals originating from fossil fuels. If this can be shown to be correct, it has a lot of implications for future human reproduction and organisms more generally. The transition from fossil fuels is not only important for overcoming climate change.

Further Reading

Bronowski J. The ascent of man. London: British Broadcasting Corporation; 1973.

Burke B. LINC complexes as regulators of meiosis. Curr Opin Cell Biol. 2018;52:22–9.

Chen HY, Jolly C, Bublys K, Marcu D, Immler S. Trade-off between somatic and germline repair in a vertebrate supports the expensive germ line hypothesis. Proc Natl Acad Sci. 2020;117:8973–9.

García-Rodríguez A, Gosálvez J, Agarwal A, Roy R, Johnston S. DNA damage and repair in human reproductive cells. Int J Mol Sci. 2018;20:31.

Ishiguro KI, Matsuura K, Tani N, Takeda N, Usuki S, et al. MEIOSIN directs the switch from mitosis to meiosis in mammalian germ cells. Dev Cell. 2020;52:429–45.

Lodish H, Berk A, Kaiser CA, Kaiser C, Krieger M, et al. Meiois. In:Molecularcellbiology.6thed.NewYork:W.H.Freeman;2008.

Nelms B, Walbot V. Defining the developmental program leading to meiosis in maize. Science. 2019;36:52–6.

Sepsi A, Schwarzacher T. Chromosome–nuclear envelope tethering–a process that orchestrates homologue pairing during plant meiosis? J Cell Sci. 2020;133:jcs243667.

Skakkebæk NE, Lindahl-Jacobsen R, Levine H, Andersson AM, Jørgensen N, et al. Environmental factors in declining human fertility. Nat Rev Endocrinol. 2022;18:139–57.

Ur SN, Corbett KD. Architecture and dynamics of meiotic chromosomes. Annu Rev Genet. 2021;255:497–526.

8

DNA and Aging

As discussed in Chap. 7, *"Generation after generation," it is apparent that in sexually producing organisms DNA transmitted from generation to generation does much better than the DNA in the mature organism, including ourselves. There is general agreement that natural selection is a declining force after DNA has been passed to the next generation. In old age, there is little selective pressure on genes that do not impact reproductive success. There are several areas of DNA and genomic biology that are important hallmarks of aging, including accumulation of mutations, shortening of telomeres at the end of chromosomes, epigenetic changes, and mitochondrial DNA damage. Given that DNA damage occurs during aging, the more efficient the DNA repair processes the greater the lifespan. While telomere shortening affects cell aging in proliferating cells (such as stem cells), it does not explain aging in nonproliferating differentiated cells. Aging is associated with progressive epigenetic change and can act as a biomarker of aging. Reactive oxygen species (ROS) in mitochondria can cause mtDNA damage, but the amount of DNA damage does not always correlate with aging. Cloning of mature cells has provided new insights into basic aging mechanisms. The involvement of specific genes has come from studies of model organisms such as C. elegans. The involvement of the insulin/IGF-1 signaling (IIS) pathway*

R. J. Rose, *DNA Know Thyself*,
https://doi.org/10.1007/978-981-97-8381-6_8

and the transcription factor FOXO3A in longevity and their relation to dietary restriction is discussed.

After transmitting our DNA to the next generation, we age and perish. However, some individuals do much better than others, some countries do much better than others, females better than males. This suggests that there is a genetic basis to aging, as well as environmental influences. While accumulated DNA damage may have a central role in aging, current thinking emphasizes the plasticity of aging and the involvement of key genes and signaling pathways in influencing aging. Experimental evidence has provided a range of insights into aging. As in many areas of biology, particularly functional genomics, it helps to look across biology and have useful biological models to help identify basic mechanisms. *Caenorhabditis elegans* is an important model for aging research. *C. elegans* is a nematode worm with a lifespan of about 3 weeks. It became a popular model for genetics and molecular biology of eukaryote tissue and organ systems based on the genetic studies of Sydney Brenner published in 1974. Nearly a decade later, Michael R. Klass identified longevity mutants of *C. elegans*, which was a boost for aging research. In addition to the genetic and molecular data available, age-related changes at the tissue level (nervous system, reproductive system, muscles), and cellular and molecular levels have been characterized. Mice and zebrafish are important vertebrate models for aging research.

There are several areas of DNA and genomic biology that have contributed to an understanding of the aging process and is a burgeoning area of research. A good place to start is with the chromosome and the chromosome ends known as telomeres, and the question of telomere attrition. Telomeres were mentioned in Chap. 2 in introducing Jack W. Szostak and his work on DNA genesis. Elizabeth H. Blackburn,

Carol W. Greider, and Szostak shared the Nobel Prize for Physiology and Medicine for "how chromosomes are protected by telomeres and the enzyme telomerase." Blackburn and Greider first identified telomerase in the single-cell eukaryote *Tetrahymena,* a ciliated protozoan. Telomerase is a conserved feature of DNA replication of chromosome ends of the linear chromosomes of eukaryotes. The way DNA of linear chromosomes is synthesized means that there must be a mechanism for maintaining the ends of chromosomes. If there wasn't such a mechanism, the chromosome would shorten after each round of DNA replication because of the way the lagging DNA strand is synthesized using Okazaki fragments (Chap. 6). Telomeres are highly repeated DNA sequences at the end of chromosomes (and bound by a protein complex called shelterin) that serve to maintain the integrity of the chromosome. One analogy is to think of them like caps at the end of shoelaces. Without those caps, the shoelaces start to fray at their ends. As long as the telomerase enzyme is active, then telomeres are maintained and there are no problems. However, somatic cells may not have, or eventually lose, telomerase activity during development. Continued shortening of telomeres eventually causes chromosome problems, leading to aging and cell death. The connection between telomeres and aging in humans is seen where diseases caused by telomerase deficiency are associated with premature aging. There is a double-edged sword associated with modifying telomere length. Increased telomere length can be associated with cancer as it favors continuous proliferation, but it is a good thing for stem cell maintenance, preventing stem cell exhaustion.

The zebrafish model (introduced in Chap. 7) is being used in telomere research and aging. Zebrafish have similar telomere and telomerase requirements to humans. Strains of zebrafish with short telomeres die prematurely compared to wild-type zebrafish. The research focus is on investigating telomere shortening and telomerase deficiency in relation to aging, cancer, and regeneration. In mice, shortened telomeres decrease lifespan while lengthened telomeres increase lifespan. In *C. elegans,* it was shown that in adults, consisting of 959 nondividing somatic cells, telomere length is not related to lifespan. Therefore, while telomere shortening affects cell aging in proliferating cells (such as stem cells), it does not explain aging in nonproliferating differentiated cells. How do shortened telomeres specifically cause aging? At a certain critical telomere length, the capping proteins (shelterin complex) don't bind properly, the DNA damage response is permanently activated, cell cycle inhibitors are induced, and cell proliferation is stopped. Cell senescence or sometimes apoptosis follow this irreversible cell cycle arrest. In differentiated cells if there is no telomere integrity, because of the inability of it to be repaired, then cellular senescence is a potential outcome. Telomere dysfunction can be linked to a range of age-related human disease pathologies of specific organs.

Genomic instability, not due to telomere dysfunction, accumulates throughout life in the form of point mutations and other DNA lesions. In a sense, DNA is always under challenge to maintain its integrity. The implications of a relationship between DNA damage and aging are that the more efficient the DNA repair process the greater the lifespan. There is experimental evidence for this. In one experiment, cell cultures were initiated from biopsies from the right hind limb from mammalian species, including humans, with different longevity. The biopsies were obtained from animals that had completed about 5% of their lifespan. The cells were subjected to UV irradiation and the amount of excision repair determined, the process to correct UV damage. The amount of repair was proportional to the logarithm of lifespan (ordered with increasing repair and lifespan—shrew, mouse, rat, hamster, cow, elephant, man). Mice deficient in DNA helicase, important in DNA repair and transcription, show premature aging. In humans, there are a group of rare genetic diseases due to DNA repair defects that mimic physiological aging. Human RecQ helicases have a pivotal role in the repair of DNA double-strand breaks. For example, defects in the RecQ helicase (WRN) encoded by the *WRN* gene cause Werner syndrome. Werner syndrome causes premature aging symptoms, and a lifespan of about 54 years. Death is usually from cancer or myocardial infarction (heart attack).

As discussed in Chap. 5, epigenetic changes are associated with changes in gene expression. Epigenetic change is associated with DNA methylation, histone modification, chromatin remodeling, and noncoding RNAs. Sometimes these changes are heritable, influencing aging in the next generation. Aging is associated with progressive epigenetic change. The DNA methylation levels and pattern, mainly cytosine 5-methylation, are altered during aging. DNA methylation is a good candidate for an age biomarker. This means that it can be used to estimate biological age as opposed to chronological age, important for understanding the aging process. Epigenetic changes are reversible, a useful characteristic for anti-aging treatments. In some respects, epigenetic change is like the accumulation of mutations, ultimately destabilizing the genome.

An interesting phenomenon with a direct bearing on aging and epigenetics is the case of induced pluripotent stem cells (iPSCs). This shows that mature cells have been reprogrammed to a younger state, a type of age reversal. The cells when initially set up expressed genes characteristic of a mature cell and then were induced to return to gene expression characteristic of an embryonic stem (ES) cell. Rather dramatically, iPSCs derived from fibroblasts from human centurions had a similar transcriptomic profile to human ES cells. Given that telomeres shorten with aging, are the induced pluripotent stem (iPS) cells limited in their potential because of the short telomeres of the donor cells? This has been investigated in mice using dermal skin fibroblasts from young and old donors. Telomeres from old donors were elongated similarly to young donors. Furthermore, iPSC telomeres had the same epigenetic marks as ES cells. Such changes to the telomeres depend on an active telomerase. It may well be that in some cases telomerase is a limiting factor and this would require transforming cells with telomerase.

A related phenomenon to iPSCs is seen in the cloning of plants and animals from mature cells. Isolated plant cells can be regenerated to produce a new seed-producing plant. Animals have been cloned by somatic cell nuclear transfer (SCNT) including from mammals as in "Dolly" the sheep. It might be expected that decreased telomere length in the donor mature cells would act to decrease life expectancy. Results from different cloned mammal species on telomere length are variable. In the case of Dolly, telomeres were 20% shorter and she died of a viral illness when 6 years old compared with the normal 12-year lifespan. Increased telomere length has been observed in cloned piglets compared to the donor fibroblasts, possibly due to higher telomerase activity in the blastocyst stage. The telomeres were of similar length in age-matched clones and normal pigs. It is possible that telomerase activity can be reprogrammed, and telomeres rebuilt in cloned animals in some cases. There are at least some cases where cloned animals have a normal lifespan. A clear picture of SCNT cloning, telomere rebuilding, and longevity of mammalian clones awaits additional and long-term data. What can be said at this point in time is that epigenetic reprogramming, including of telomeres, is feasible and provides an approach to reduce aging by understanding the reprogramming process. This does not however overcome unpredictable DNA mutations.

What has been considered so far has focused on the nuclear genome and the chromosomes of the nucleus and how they influence the genetics and epigenetics of aging. The mitochondrion contains DNA that codes for only a few genes and it imports most of its proteins. As the powerhouse of the cell, it is a major contributor to life processes and a focus for aging research. As for nuclear DNA, mitochondrial DNA accumulates mutations during aging. Point mutations and deletions occur in mitochondria in aging humans, monkeys, and rodents, with varying severity in different tissues, including the heart, muscle, and brain. Do these changes cause aging or are they just correlated with the aging process? This question has been addressed experimentally in mice. Trifunovic and coworkers genetically engineered mice to produce a mutated mitochondrial DNA polymerase-γ that eliminated its proofreading capacity. This error-prone polymerase caused extensive mitochondrial DNA mutations, and the mice showed reduced lifespan and premature aging phenotypes such as weight loss, hair loss, osteoporosis, and heart enlargement. The mutated animals also showed decreased activity of enzymes involved in the respiratory chain and ATP (the main energy store) production. There was also an increase in reactive oxygen species (ROS), which is a by-product of ATP formation and can cause oxidative damage. These data are consistent with the mitochondrial-free radical theory of aging (MFRTA) where aging is caused by defects in mitochondria, leading to increased ROS causing mtDNA mutation. Excessive ROS can cause DNA and cellular damage. There is no doubt that ROS can be destructive at high levels. There is interesting recent data showing some cells can actively reduce ROS to enhance cellular integrity. Humans (and other mammals) have a reservoir of immature oocytes (eggs) at birth that need to be stored in excellent condition for more than 40 years, given they are the gametes for the next generation. A strategy is used to reduce ROS in primordial eggs by modifying the flow of electrons in the mitochondrial electron transport chain by repressing complex I assembly. Though there is evidence for MFRTA being important in certain contexts, it does not provide an explanation for aging that is generally accepted. MFRTA as a conceptual framework for the role of mtDNA as a primary driver of aging continues to be debated, in part this is because the number of mtDNA mutations observed in normal aging is not consistently high. The role of ROS is complex as while it can cause cellular damage, in some situations it acts as a signal to initiate positive responses. Mitochondria have many roles in the cell in addition to energy production that can influence aging. Mitochondria have a key role in activating programmed cell death to remove damaged cells, known as apoptosis. It is important that mitochondria quality be maintained for healthy aging. Mitochondrial quality is maintained by mitochondrial fission and fusion and the removal of damaged mitochondria by mitophagy.

Unsurprisingly, aging research is predominantly directed at animals with a view to understanding the human aging process. However, insights can be derived from plants and bacteria. Some plants can live for thousands of years while others have a predetermined annual cycle. While in some cases this is by continual cloning of the original plant, plant stem cells can continue to produce new cells for differentiation for long periods of time without exhausting the stem cell's capacity. This argues for the immortal capacity of DNA in a nonsexual somatic cell process. Some authors have suggested that plant stem cells in meristems act more like germline cells. Having active shoot apical meristems means that aged leaves can die and be replaced, essentially organ replacement.

An interesting bacterial experiment was initiated in 1988 and continues to this day, primarily to analyze evolution in *Escherichia coli*. Twelve flasks were set up by Richard Lenski at Michigan State University, and the cultures continue to this day. This represents 75,000 generations of growth. These 12 different lines of a prokaryotic cell show generation after generation that their DNA is essentially pristine. However, evolutionary change has occurred with one line evolved to have the ability to consume citrate instead of just glucose. This is classic DNA, stable but with a capacity for evolutionary change.

What has been considered so far are four key hallmarks of aging focusing on DNA. These are DNA and genome instability, telomere attrition, epigenetic alterations and epigenetic reprogramming, mitochondrial DNA and mitochondria dysfunction. Lopez-Otin and coauthors have suggested five additional hallmarks that can be linked back to the fundamental latter four. The additional five are stem cell exhaustion, proteostasis (protein homeostasis), cellular senescence, deregulated nutrient sensing, and altered intercellular communication (immunological, hormonal). Stem cell exhaustion, proteostasis, cellular senescence, and deregulated nutrient sensing connect back to DNA damage.

Given then that there is a broad understanding of major contributors to aging, how is this being translated in terms of increasing lifespans? Experiments can be turned over quite quickly in the *C. elegans* aging model and has had a major influence on aging research. An influential discovery in *C. elegans* occurred in 1993 by Cynthia Kenyon and coworkers, which showed that a mutation in the *daf-2* gene doubled the lifespan of *C. elegans* and remains an active area of research. The *daf-2* gene is a regulatory gene encoding an insulin/IGF-1 (insulin/insulin growth factor-1) receptor for the insulin/IGF-1 signaling (IIS) pathway. The IIS pathway is an evolutionary conserved aging regulatory pathway. When the activity of the IIS pathway is downregulated (as with the *daf-2* mutation), the transcription factor DAF-16 (FOX0 in mammals) is upregulated, leading to reduced age-related degenerative change, slowing of metabolism, and increased longevity. These studies have been used to posit the thesis that aging is an active process involving specific regulatory pathways. There is evidence that aging in humans is also influenced by proteins activated by the downregulated IIS pathway as in *C. elegans*. Longevity in humans has been shown to be associated with the regulation by the FOXO transcription factor FOXO3A. Centenarians were found to have unique variations in FOXO3A.

Dietary restriction (DR) of 40–60% (without malnutrition) can delay aging and increase lifespan in *C. elegans*, *Drosophila*, and mice. A contributor to the dietary effects is a downregulated IIS pathway, similar to what occurs with *daf-2* mutants. Another contributor to the diet response is the mTOR (target of rapamycin) signaling pathway, which acts as a switch between growth and maintenance. In *C. elegans*, mutants for the mTOR and IIS pathways have a nearly fivefold increase in lifespan. A third contributor are certain sertuin proteins that are NAD^+ (a metabolically important redox coenzyme) dependent and act as epigenetic regulators by modifying histone proteins. Decreased activity of the sertuin protein SIRT1 has been linked to increased DNA damage, reduced mitochondria quality, and downregulated antioxidant defense. Dietary restrictions as a general strategy would seem an unlikely way to activate specific genes and pathways that are important for normal metabolism. It has been suggested that this DR response is a characteristic of stress responses, shifting the physiology of the organism to cellular protection and maintenance.

The research on DR has been carried out on model organisms, and there is now an emphasis on translating this understanding to humans. While the identification of key genes has enhanced the understanding of aging, intervention pharmacologically has had mixed success. Metformin is widely used in the treatment of type 2 diabetes. However, metformin effects include decreased IIS and mTOR signaling, reduced ROS and DNA damage—age-related factors. The value of metformin in aging is still an open question, with clinical trials underway. Rapamycin, the mTOR pathway inhibitor, has been shown to inhibit aging in yeast, *C. elegans*, *Drosophila,* and mice. This drug has been approved for human use for various therapies including cancer. Clinical trials have shown an improvement in the physiology of aging-related diseases. Researchers are optimistic about the potential use of rapamycin in antiaging therapy. Sertuin activators have had mixed success in clinical trials. Exercise enhances healthy aging and there is evidence for increased lifespan, but the molecular basis is unclear. Diet remains a widely discussed topic in the community but with no clear science in relation to lifespan, though epidemiological data from human populations is consistent with DR studies in model organisms. DR studies are being advanced by testing different types of fasting strategies and relating these to the cell and molecular biology of aging.

At the cell level, there are actively dividing adult stem cells as well as differentiated cells performing specific functions to be considered in aging research. Aging in relation to DNA damage is also part of the pursuit of the understanding of stem cells and cancer, particularly in the context of stem cell exhaustion and diseases of aging. Research on DR and aging is revealing specific genes that connect to aging. Aging research is now integrating the hallmarks of aging to develop evidence-based practical applications to healthy aging and longevity. Research progress will be interesting!

Further Reading

Campisi J, Kapahi P, Lithgow GJ, Melov S, Newman JC, Verdin E. From discoveries in ageing research to therapeutics for healthy ageing. Nature. 2019;571:183–92.

Green CL, Lamming DW, Fontana L. Molecular mechanisms of dietary restriction promoting health and longevity. Nat Rev Mol Cell Biol. 2022;23:56–73.

Greider CW, Blackburn EH. Identification of a specific telomere terminal transferase activity in *Tetrahymena* extracts. Cell. 1985;43:405–13.

Hart RW, Setlow RB. Correlation between deoxyribonucleic acid excision-repair and lifespan in a number of mammalian species. Proc Natl Acad Sci U S A. 1974;71:2169–73.

Kenyon CJ. The genetics of ageing. Nature. 2010;464:504–12.

López-Otín C, Blasco MA, Partridge L, Serrano M, Kroemer G. The hallmarks of aging. Cell. 2013;153:1194–217.

Marion RM, Strati K, Li H, Tejera A, Schoeftner S, Ortega S, Serrano M, Blasco MA. Telomeres acquire embryonic stem cell characteristics in induced pluripotent stem cells. Cell Stem Cell. 2009;4:141–54.

Schumacher B, Pothof J, Vijg J, Hoeijmakers JH. The central role of DNA damage in the ageing process. Nature. 2021;592:695–703.

Trifunovic A, Wredenberg A, Falkenberg M, Spelbrink JN, Rovio AT, et al. Premature ageing in mice expressing defective mitochondrial DNA polymerase. Nature. 2004;429:417–23.

Zhang W, Qu J, Liu GH, Belmonte JC. The ageing epigenome and its rejuvenation. Nat Rev Mol Cell Biol. 2020;21:137–50.

9

RNA and DNA Viruses Reproduce in Cells

With the evolution of cells, carriers of DNA, organisms came to occupy all types of environmental niches. There is another niche—the cell itself. Here, DNA and RNA organisms live within cells to reproduce. These are the viruses. Viruses typically consist of a protein coat (nucleocapsid), encasing nucleic acid, RNA or DNA, with a membrane envelope with embedded proteins for enveloped viruses. Viruses are classified according to the type of nucleic acid and the way it is replicated in the cell. SARS-Cov-2 virus is an example of a (+) strand RNA virus. An example of a DNA virus is the monkeypox virus, a member of the poxvirus family. Another group of viruses with quite a different life cycle are the RNA (+) strand retroviruses, which replicate via *a DNA intermediate. An example is the human immunodeficiency virus (HIV), which causes the acquired immunodeficiency syndrome (AIDS). A final example of an important human virus is the Ebola virus, which infects humans and other primates. This virus is an example of an RNA (-) strand virus. The plant virus considered is the alfalfa mosaic virus (AMV). All the four human viruses considered are enveloped viruses and have broad similarities in their life cycles in terms of entry, replication, and exit from the cell. However, their different nucleic acid biology provides important differences. Knowing the life cycle of the viruses provides the background for the therapeutic approach to the control of these*

R. J. Rose, *DNA Know Thyself*,
https://doi.org/10.1007/978-981-97-8381-6_9

serious human viruses. The therapeutic approaches considered are via *vaccination and antivirals. For the plant, AMV virus genetic engineering methods are described.*

All cellular organisms can be infected with viruses, but in this chapter the focus is on eukaryotes, plants, animals, and humans. Bacteriophages are not considered here, but are an important group for molecular biology studies, and research continues on their potential as therapeutic agents for controlling bacterial infection. In viruses infecting eukaryotes, a viral protein (e.g., spike protein of SARS-Cov-2) recognizes the cell to be infected and enters the cell where it reproduces, then exits the cell and the mature virus is ready to infect another cell.

Viruses are classified according to the type of nucleic acid and the way it is replicated in the cell. This classification system basically comes from a paper written by David Baltimore in 1971. David Baltimore won the Nobel Prize (with Renato Dulbecco and Howard Temin) in 1975 for his research on RNA viruses that could make DNA from RNA using a reverse transcriptase. This was not only important in virus research but in molecular biology generally. Making RNA from DNA meant that the central dogma of molecular biology where DNA → RNA, with the RNA translated into protein, was not the only way to go, RNA could be the template for DNA. In this chapter, some different types of well-known viruses will be considered to show their life cycles and the different ways they reproduce. Approaches to control the infection of these viruses are also considered.

Right now, the most famous (or infamous) virus is the coronavirus SARS-Cov-2, which causes COVID-19. The SARS-Cov-2 virus has a positive sense single-strand RNA (+) genome. This means that in an infected cell the RNA can act directly as an mRNA molecule to be translated into viral proteins. The genomic (+) strand RNA serves as a template for the synthesis of (-) strand RNA using an enzyme called RNA-dependent RNA polymerase (RdRp). The (-) strand RNA then serves as a template for more (+) strand RNA synthesis. This (+) RNA can act as mRNA for the synthesis of more proteins or it can serve as the replicated genome for the virus. The RNA (+) genome is packaged with the help of a nucleocapsid (designated N) RNA-binding protein. The genome is encased in a membrane envelope with embedded proteins, designated E (envelope protein), M (membrane protein), and S (spike protein). The 24–40 protruding S proteins are randomly arranged around the spherical virus. The spike proteins are coated with glycans, complexes built up from sugars.

The first step in SARS-Cov-2 infection is the binding of the receptor binding domain of the SARS-Cov-2 spike protein to the ACE2 receptor on the outside of human throat and lung cells, and the virus can be distributed to other internal organs via the blood. After binding to the receptor, the virus membrane fuses with the cell membrane, forming a pore to enable the viral RNA to be released into the cytoplasm. The virus RNA literally takes over the cell. Viral proteins encoded in the viral RNA are produced and can suppress the translation of the host's cell mRNA while allowing the translation of viral RNA. Further, cellular RNAs are prevented from exiting the nucleus, so they cannot be used to induce a defense. The viral RNA is ultimately translated to produce the proteins necessary for its reproduction. The membranes of the endoplasmic reticulum (a cytoplasmic membrane system) are co-opted to surround the viral RNA in vesicles. Sequestered within the vesicles, viral RNA replication and more translation occur for assembling new virus. The newly assembled virus exits the cell via the Golgi apparatus, a specialized organelle that usually packages proteins in vesicles for various destinations in the cell or for secretion from the cell. The virus is then free to infect another cell.

Some viruses have a DNA genome. An example is the monkeypox virus (MPOX), which can infect humans and has recently drawn more concern. The MPOX virus is closely related to the smallpox virus. It is an enveloped virus, with a relatively large (approximately 197 kb) double-stranded DNA genome. This contrasts with the RNA (+) SARS-Cov-2, which is approximately 30 kb. MPOX is a member of the poxvirus family where most detailed studies have been carried out on the prototype poxvirus, Vaccinia. Poxviruses, as DNA viruses, have a lower mutation rate than RNA viruses; however, the large genome can undergo structural changes, resulting in gene loss or gain. Associated with the membrane envelope are 20 or more viral proteins, with several proteins required for the entry of the virus into the cell. The virus enters the cell via fusion with the cell membrane, releasing the viral DNA into the cell through the pore produced by the membrane fusion. The DNA codes for approximately 200 proteins, including proteins required for genome replication, genome transcription, and virus assembly. The replication of the DNA and assembly of the virus occurs within membrane vesicles initiated from the rough (refers to bound ribosomes) endoplasmic reticulum. The fully assembled replicated virion (i.e., the capsid encased DNA of the virus) exits via a Golgi apparatus pathway and obtains two additional membrane envelopes. The fully enveloped virus is transported with the assistance of the cytoskeleton to the cell membrane. The fusion to the cell membrane and exocytosis follows, to complete the life cycle and be ready for a new round of infection.

Another group of viruses with quite a different life cycle are the RNA (+) strand retroviruses, which replicate via a

DNA intermediate. An example is the human immunodeficiency virus (HIV), which causes the acquired immunodeficiency syndrome (AIDS). DNA is made from RNA using a reverse transcriptase enzyme, which as mentioned earlier was an important biological discovery for molecular biology. Similar to the other two viruses discussed, it is bounded by a membrane envelope with embedded proteins but its replication and assembly are quite different. A viral transmembrane protein (ENV) with two parts, a gp120 and a gp41 subunit, binds to the receptor CD4. This is followed by binding to a cellular co-receptor, cell membrane fusion, formation of a fusion pore, and entry of the nucleocapsid encased RNA into the cell. The $CD4^+$ T cells of the immune system are preferentially infected. The RNA is released from its nucleocapsid protein coating where a reverse transcriptase using the RNA template is able to produce DNA. What then happens to the viral DNA is fascinating, albeit making control of the virus difficult, which becomes essentially "hidden."

Once the viral DNA is formed, it enters the nucleus and another enzyme comes into play. This is the HIV integrase that splices the viral DNA into the host's genome. The virus when integrated into the host DNA is known as a provirus. If the $CD4^+$T cell is then activated, the viral DNA is transcribed to mRNA. This mRNA serves as the replicated RNA (+) viral genome and is also translated to produce proteins in the cytoplasm required (after trimming with a protease) for the assembly of the virus. The HIV Gag protein is an important protein for virus assembly and release. The virus exits from the cell primarily via a budding process involving the cell membrane forming the envelope of the virus with its associated proteins. If the $CD4^+$T cell is not activated, the provirus remains until this activation occurs, meaning that the virus in the provirus form can remain indefinitely and needs to be considered in treating AIDS.

A final example of an important human virus is the Ebola virus, which infects humans and other primates. The Ebola virus causes hemorrhagic fever, which has an extremely high fatality rate. Ebola is an example of a negative strand RNA virus having an RNA (-) genome. This means that it is of opposite polarity to mRNA. The Ebola virus is a filamentous enveloped virus with the RNA (-) genome packaged within a helical nucleocapsid surrounded by a membrane with embedded and associated proteins. Viral membrane components important in binding to cell membranes and subsequent uptake are two glycoproteins (GP1 and GP2) and phosphatidyl serine. The cell receptors, which are not highly specific, are lectins and phosphatidyl serine receptors that interact with the viral glycoproteins and viral phosphatidyl serine, respectively. The virus is then internalized by cell membrane protrusions wrapping around the virus forming a membrane-bound endosome. Exit of the virus from the endosome is facilitated by the viral envelope fusing with the endosomal membrane in a way that releases the viral nucleocapsid plus RNA (-) and associated proteins. Once in the cytoplasm, the RNA (-) is freed from the nucleocapsid.

The proteins brought into the cell with the virus facilitate primary transcription to produce mRNA from the RNA (-) and the cellular machinery carries out translation. The newly translated proteins then facilitate secondary transcription and viral replication. The viral enzyme RNA-dependent RNA polymerase (RdRp) complex is key for the transcription and replication of the RNA (-) genome. Viral replication involves an RNA (+) intermediate, with the RNA (+) then serving as a template for viral RNA (-) genomes. Following replication, there is nucleocapsid packaging followed by transport via the cytoskeleton to the cell membrane where budding occurs. At budding sites, the virus protein VP40 enables the completion of virus morphogenesis in concert with bud formation. The virus particle is finally formed as a membrane-bound nucleocapsid encasing the RNA (-) genome.

So far examples of the life cycle of viruses that infect humans (and also some animals) have been considered. Plants also have viruses, and they have some similarities to animal viruses. An example is the AMV virus (alfalfa mosaic virus) that infects a wide range of species, including alfalfa (*Medicago sativa*) and the model legume *Medicago truncatula*. The virus is not enveloped, as are most plant viruses, unlike the viruses previously discussed. The AMV virus is generally described as bacilliform in shape (short rods) and consists of RNAs packaged as a nucleocapsid with a coat protein. The AMV viruses have an RNA (+) genome that is multipartite with four components. RNA1 and RNA2 code for proteins involved in RNA replication, RNA3 codes for a movement protein involved in cell-to-cell movement together with a copy of the coat protein. While RNA4 encodes a separate copy of the coat protein. Each of the four RNAs are separately encapsidated with the coat protein and assembled into the intact virus. The virus can come in the seed or via aphids. The virus is able to spread from cell to cell via plasmodesmata with the assistance of the movement and coat protein encoded by RNA3. Plasmodesmata are highly structured cytoplasmic connections between plant cells. The RNA genome is replicated with the RdRp enzyme, with the coat protein also contributing to the replication process.

Knowledge of DNA and RNA replication of particular viruses has led to understanding of the life cycle of the virus. Each virus has its own peculiarities, and a lot of research is invested in the understanding involved. The hope is that this understanding will enable targeting of the replication of the virus. The mainstay of virus control in humans has been via vaccination, which has proven successful. The utilization of the immune system and vaccination for virus control dates back to 1796 and Edward Jenner's successful work on the use of cowpox inoculum against smallpox. Successful vaccines have been developed for many viruses other than smallpox, for example, polio, measles, mumps, rubella, hepatitis A and B, and influenza. Vaccines act through B and T cell lymphocytes. Specific antigens (virus proteins in the vaccine) cause B cells of the immune system to produce antibodies that recognize the antigen and act to eliminate the virus (antibody immunity). Further, T cells can produce antigen-specific surface receptors that recognize and kill infected cells (cell-mediated immunity). After vaccination, the memory of B and T cells of the virus antigens activates the immune system to eliminate subsequent viral infections.

Improved knowledge of DNA and RNA biology has led to additional strategies for virus control. In the second part of this chapter, I wanted to consider the strategies used against the five viruses considered, starting with SARS-Cov-2. The first successful mRNA vaccines were developed for SARS-Cov-2. Prior to the mRNA vaccine, successful immunization utilized inoculation with attenuated or inactivated virus, and more recently with purified or recombinant proteins. The mRNA vaccines can be produced much more quickly, enabling a rapid response to variants. Attempts to deliver mRNA into cells to produce proteins probably date to early experiments in 1989 by Robert Malone. The discovery of mRNA as discussed in Chap. 4 was in the 1960s. Another important development was the ability to synthesize mRNA molecules without the need to isolate the mRNA from cells. Besides the need to reproducibly introduce intact mRNA molecules into cells, there was the need to improve the stability of the introduced mRNA. Isolated mRNA is notoriously unstable. In general terms, it took numerous scientists over 30 years to have mRNA vaccine success. However, at the end point, a few individuals and their teams made the final steps, though attribution is never straightforward. One of the problems with mRNA is that it stimulated an early immune response, leading to an inflammatory reaction. Researchers Katalina Kariko, Drew Weissman, and coworkers in 2005 found that by modifying the uridine nucleotide to a pseudouridine, the inflammatory response could be eliminated. This is used in the BioNTech/Pfizer and Moderna vaccines. Kariko and Weissman received the Nobel Prize in 2023 for their uridine modification work. For the delivery of the modified RNA, liposome vesicle-like structures were developed further into lipid nanoparticles. These are essentially modified membrane vesicles with specially defined lipids that house the mRNA. Pieter Cullis was an important pioneer in this area. Biotech startups were important in the development of these technologies. Özlem Türecki and Uğur Şahin cofounded the mRNA vaccine firm BioNTech and used mRNA in lipid nanoparticles, then with their help Pfizer developed the successful mRNA vaccine using the spike protein mRNA for the antigen. The Moderna vaccine was similarly developed. These mRNA vaccines have been successful in preventing serious illness and death for many with COVID-19. Another important vaccine utilized DNA rather than RNA to code for the spike protein and used a virus as a delivery system rather than lipid nanoparticles. Sarah Gilbert and coworkers at the University of Oxford, together with AstraZeneca, used a replication-deficient adenovirus containing the full-length DNA sequence of the spike protein. The Pfizer and Moderna mRNA vaccines have become the vaccine of choice, consistent with available data, for example, Rotshild et al. in a meta-analysis reported (*Scientific Reports* 2021) that mRNA vaccines had the highest efficacy in preventing symptomatic COVID-19 compared to other vaccines. Phase 3 clinical trials reported by Pfizer and Moderna showed about 90% efficacy in preventing COVID-19. With the experience of mass vaccination, the effect of factors such as age, underlying medical conditions, number of doses, and length of the immunity is now more available for clinicians and government policy.

The mRNA vaccines emphasize the value of DNA-related technology in disease prevention. Another ancillary approach to COVID-19 disease is the use of antivirals. The antivirals target processes that occur in the virus life cycle to prevent its replication. The oral antiviral molnupiravir inhibits the enzyme RdRp preventing replication of the RNA (+) of the virus. Another antiviral is Paxlovid®, which is a protease inhibitor. Proteases are required to cleave precursor proteins to form mature proteins required for replication and assembly of the virus.

Modern vaccines for the control of MPOX viral disease use a highly attenuated nonreplicating Vaccinia virus. There is limited data on two antivirals that act on different points of the life cycle. The antiviral inhibitor of DNA replication brincidofovir had no convincing clinical effects. Tecovirimat inhibits release of the virus from the cell by inhibiting a protein involved in forming the viral envelope. This antiviral caused a shorter duration of symptoms in a limited observational study with small numbers and is regarded as promising.

Despite extensive efforts, an effective vaccine against HIV has not yet been developed. According to Barton Haynes (Duke University) and coauthors, writing in *Nature Reviews Immunology*, this is due to the "extraordinary genetic diversity of HIV-1 and its complex mechanisms of immune evasion." However, antiretroviral therapy (ART) has been very effective in the treatment of HIV, such that near-normal lifespans are obtained. The antiviral inhibitors target key points in the life cycle of the virus. First are the reverse transcriptase inhibitors that prevent DNA formation from the RNA (+) template. Three reverse transcriptase inhibitors commonly used at different stages of treatment are tenofovir alafenamide, lamivudine (nucleoside inhibitors), and rilpivirine (non-nucleoside inhibitor). The other type of inhibitors used in ART target the integrase that incorporates the DNA into the nuclear genome of the host cell. Two commonly used integrase inhibitors are bictegravir and cabotegravir.

The most advanced vaccines against the Ebola virus utilize a harmless surrogate virus as a vector. The vaccine known as ERVEBO® or rVSV-EBOV acts against the Zaire Ebolavirus strain. It has proven clinical efficacy and US FDA approval. The vaccine has the vesicular stomatitis virus (VSV) as the backbone, which carries the Ebola-GP gene. This gene codes for the Ebola glycoprotein required for the attachment of Ebola to host cells, which is the antigen producing the immunity. The VSV virus allows entry of the vaccine into the cell like the AstraZeneca COVID-19 vaccine strategy, which uses the adenovirus. Research on an mRNA vaccine is underway. As for the other three viruses discussed, antivirals are being developed. The most promising are small-molecule inhibitors of the RNA-dependent RNA polymerase L required for viral replication.

Viruses are also important plant pathogens where DNA-related technology is important in the contemporary approaches to control. Locating resistant plant strains for plant breeding and destruction of infected plants, together with biosecurity, were the only previous approaches. Plants do not have circulating cells involved in the human immune system to facilitate immunization. Therefore, genes to inhibit the replication of the virus need to be transferred into the genome. The AMV virus discussed here can be made immune to the disease via producing transgenic plants with the coat protein gene (see also Chap. 11). While coat-protein mediated resistance works, the mechanism is variable. In cases where the amount of resistance correlates with the amount of the coat protein, the resistance is likely due to the inability of the virus to release the viral RNA from the nucleocapsid due to excess coat protein. In other cases, the resistance is independent of coat protein quantity and is likely due to silencing of the viral coat protein gene transcripts due to the production of double-stranded RNA (micro RNAs, Chaps. 5 and 11), which degrades the viral transcript. The different mechanisms are due to the way the number and orientation of coat protein genes are inserted into the genome. The silencing of viral gene transcripts is now a new approach to plant virus resistance research using diverse genes.

The four human viruses discussed are all enveloped RNA or DNA viruses that have their unique structure and life cycle. This knowledge is critically important in determining the protein antigens used in vaccination and in the development of antivirals. The success of antivirals for HIV infection control is illustrative of the significance of antivirals. In other cases, such as with COVID-19, vaccination and antivirals are important in the control of the disease. Antivirals targeted to vulnerable parts of the virus life cycle are a fertile area for future research. It is interesting that the molecular biology of antivirals has overlaps with cancer biology (and other areas) where blocking the activity of key genes provides an important approach to disease control.

Further Reading

Baltimore D. Expression of animal virus genomes. Bacteriol Rev. 1971;35:235–41.

Cihlar T, Fordyce M. Current status and prospects of HIV treatment. Curr Opin Virol. 2016;18:50–6.

Dolgin E. The tangled history of mRNA vaccines. Nature. 2021;597:318–24.

Hoenen T, Groseth A, Feldmann H. Therapeutic strategies to target the Ebola virus life cycle. Nat Rev. Microbiol. 2019;17:593–606.

Karikó K, Buckstein M, Ni H, Weissman D. Suppression of RNA recognition by toll-like receptors: the impact of nucleoside modification and the evolutionary origin of RNA. Immunity. 2005;23:165–75.

Klimas N, Koneru AO, Fletcher MA. Overview of HIV. Psychosom Med. 2008;70:523–30.

Lindbo JA, Falk BW. The impact of "coat protein-mediated virus resistance" in applied plant pathology and basic research. Phytopathology. 2017;107:624–34.

Schramm B, Locker JK. Cytoplasmic organization of POXvirus DNA replication. Traffic. 2005;6:839–46.

Scudellari M. How the coronavirus infects our cells. Nature. 2021;595:640–4.

Xiang Y, White A. Monkeypox virus emerges from the shadow of its more infamous cousin: family biology matters. Emerg Microb Infect. 2022;11:1768–77.

10

DNA and the Rise of Modern Biotechnology

Biotechnology utilizes microorganisms, plant, and animal cells to produce industrial products, therapeutic proteins, and research tools. It also includes gene therapy and genetic engineering. Traditionally, biotechnology products include alcoholic beverages such as wine and beer involving fermentation driven by yeast, cheese using bacteria, and bread using yeast. Modern biotechnology is associated with the understanding of the structure of DNA, its replication and translation, the proteins produced, and the way DNA can be manipulated. The beginning of modern biotechnology was the cloning of a piece of DNA cut with restriction enzymes and spliced into plasmids in E. coli in 1973. This cloning technique led to the production of human insulin and human growth hormone, followed by other therapeutic proteins. New techniques allowed the discovery and isolation of specific genes of interest. Techniques included nucleic acid hybridization, the reverse transcriptase enzyme (RNA→DNA), the polymerase chain reaction (PCR), and continued improvements in sequencing. The development of the US National Center for Biotechnology Information (NCBI) database was central to gene discovery and continues to have an important role. The discovery and cloning of the BREAST CANCER GENE 1 (BRCA1) was an important step in the isolation of diverse genes. The completion of the draft DNA sequence of the human genome in 2001 marked an

R. J. Rose, *DNA Know Thyself*,
https://doi.org/10.1007/978-981-97-8381-6_10

important milestone. The development of new-generation sequencing led to the sequencing of the genome of many organisms and the routine sequencing of human genomes for genetic analysis. For human gene therapy, it was necessary to develop suitable vectors. Replication-deficient recombinant adenovirus that enters the nucleus and is not integrated into the genome and adeno-associated virus vectors (AAV) is discussed. Genetic engineering (GE) of plants involving gene insertion is possible using Agrobacterium tumefaciens-based vectors. GE cattle produced by somatic cell nuclear transfer (SCNT) have become part of breeding strategies.

The demonstration of recombinant DNA molecules in 1972 by Paul Berg and coworkers was an important step in DNA biology and biotechnology (see Chap. 4). The paper demonstrating the splicing of a DNA piece into the DNA of bacterial plasmids, by Stanley Coen, Herbert Boyer, and coworkers in 1973, is commonly considered the beginning of DNA-related biotechnology. A DNA molecule can be cut at a specific site with a specific restriction enzyme (e.g., *EcoR1*) and a foreign piece of DNA (with compatible ends generated by the same enzyme, *EcoR1*) inserted at that site. The joining of the new foreign DNA piece to the cut DNA molecule with the compatible ends is assisted by a joining enzyme called DNA ligase.

The Coen–Boyer work was carried out with bacterial plasmids, circular pieces of DNA that replicate independently, but in concert, with the bacterial chromosome. DNA cut from one plasmid with EcoR1 was spliced into another plasmid cut with *EcoR1* to form a novel plasmid, which was transformed into *Escherichia coli* where it was replicated. A method for cloning DNA fragments had been developed. Subsequent work showed that DNA from prokaryotes and eukaryotes could be cloned in recombinant plasmids in *E. coli*. Coen and Boyer, through their universities (Stanford and University of California, San Francisco), controversially but successfully patented this DNA cloning technique using recombinant DNA (approved by the US Patent Office in 1980). The patent was commercialized by nonexclusive licensing to many companies. The patent expired in 1997. The first biotechnology company based on DNA technology was Genentech, set up by Boyer and Robert Swanson (a venture capitalist) in 1976. Paul Berg and Janet Mertz, reflecting on the onset of the DNA technology era, wrote that "the emergence of the biotechnology industry followed naturally from the encouragement of academic scientists to patent their research discoveries and to explore their newly discovered entrepreneurial instincts." Herbert Boyer marked a change in the way science was commercialized from a university and the perception of biotechnology. Boyer did not get a Nobel Prize (though arguably deserving), but his development of Genentech (with Robert Swanson) led to him being featured on the cover of *Time* magazine in 1981. Biotechnology startups have continued to be part of the academic landscape until the present day. In the current era, BioNTech has played a pivotal role (see Chap. 9) in the development of COVID-19 vaccines.

Recombinant DNA technology allowed therapeutic proteins to be produced in commercial quantities. Early products were insulin and human growth hormone (HGH). The inability to synthesize insulin in the beta cells of the pancreas causes type 1 diabetes due to elevated blood glucose as insulin is required to stimulate uptake of glucose into body cells. Insulin was initially obtained from the pancreas of pigs and cows. The first human recombinant DNA insulin was produced in 1978 by David Goeddel and coworkers from Genentech, and later in 1978 Genentech and Lilly signed an agreement to commercialize human insulin.

Biosynthetic human insulin using recombinant DNA technology was first marketed in 1982 as Humulin®.

Human growth hormone (HGH) is produced in the pituitary gland located at the base of the brain. A deficiency in HGH in childhood leads to dwarfism. In the past, clinical use of HGH for HGH deficiency was obtained from human cadavers. Sadly, some HGH from cadavers were contaminated with Creutzfeldt–Jakob disease (CJD, a rare degenerative brain disease similar to mad cow disease), causing this source of treatment to be discontinued. Thankfully recombinant DNA technology for HGH production became available. Genentech produced and marketed HGH in 1985 (known as Protropin) by using recombinant DNA technology in a bacterial system using the human gene. Several other Biotech companies quickly followed with their own HGH product.

To produce a therapeutic protein using recombinant DNA technology, the gene sequence needs to be known. If the protein has been purified and the amino acid sequence obtained, then the DNA sequence can be determined then chemically synthesized, as was done for insulin. In the years between the first therapeutic proteins and the publication of the draft human genome in 2001, it was necessary to isolate the sequence of the protein of interest from the unsequenced genome. One common way to isolate a specific gene is to use what is known as cDNA libraries. Messenger RNA (mRNA) is isolated from a tissue that synthesizes the gene of interest, but there are of course other genes being transcribed. The total mRNA is reverse transcribed by the reverse transcriptase enzyme to produce the more stable DNA. This cDNA is inserted into plasmids and the plasmids into bacteria. The end result is a library of bacterial colonies growing on agar. One (or some) of these colonies contains the cDNA sequences of interest. This is detected by using a labeled probe (e.g., radioactively labeled) that will base pair (hybridize) to the target gene sequence that has been denatured to form single strands. The probe is key, and the availability of a suitable probe will depend on what is known of the gene of interest. If some stretches of amino acids are known for its protein, then the oligonucleotides can be deduced by working backward and chemically synthesized. There may be some partial DNA sequence available from which a probe can be obtained. Biological research involves many organisms and information on human genes and their protein can be obtained from genes of other organisms that show some homology with the human gene. Importantly, in relation to the latter points, was the development and funding by the United States of the National Center for Biotechnology Information (NCBI) database. NCBI is part of the US National Library of Medicine, a branch of the National Institutes of Health, and was established in November 1988. The NCBI database increased in sophistication year by year to the present day. From this database, information can be obtained on the contemporary knowledge of a given gene. The use of cDNA allows the gene sequence consisting of all the exons to be determined. If the introns (as well as the coding exons) need to be determined, then DNA libraries, rather than cDNA libraries, need to be used where the DNA is cut with restriction enzymes.

In 1985, a very important discovery was made that greatly assisted gene discovery and DNA technology/biotechnology more broadly. The discovery of the polymerase chain reaction (PCR) was made by Kary Mullis while working on oligonucleotide probes at the Cetus Corporation in California. He received the Nobel Prize in Chemistry in 1993. He had an unusual and controversial career, not your usual Nobel Prize-winning scientist. There is no doubt about the significance of his discovery. Today, worldwide the "gold standard" PCR test for COVID-19 is a household word. The PCR method allows a tiny amount of DNA to be copied in large quantities over a short period. How does it work?

In PCR, the two antiparallel strands of the double-stranded DNA are separated by denaturing the DNA using heat. At a lower temperature, primers (a short stretch of 20–30 nucleotides) base pair at opposite ends of the target DNA. The DNA polymerase then extends each strand by adding nucleotides (from 3′ end of the primers) that base pair along each strand. This gives two replicated DNA molecules. By applying heat again, the DNA molecule's two strands are separated and the process repeats, giving four DNA molecules and so on. Starting with a single DNA molecule, 20 PCR cycles give more than one million copies (2^{20}). A PCR machine known as a thermal cycler can cycle the temperature requirements for each PCR cycle. An important refinement of PCR was the use of the extremely heat-resistant DNA polymerase called *Taq* polymerase isolated from the bacterium *Thermus aquaticus*. *T. aquaticus*, a member of the Archaea (Chap. 2), thrives in hot springs and was discovered in the thermal springs of Yellowstone National Park. The discovery and cultivation of *T. aquaticus* were reported by Thomas Brock and Hudson Freeze, another discovery linking fundamental research to biotechnology.

In the 1990s, gene discovery accelerated using the increasing number of molecular biology tools and increasing information on the NCBI database, but it was still a difficult process. The human genome project commenced in 1990 and the draft completed in 2001. An important step in gene discovery and cloning was the discovery and sequencing of the *BREAST CANCER GENE 1 (BRCA1)*.

This gene was found in 1990 to be located on a locus on human chromosome 17 by a group led by Mary-Claire King at the University of California, Berkeley. With this knowledge, a group led by Mark Skolnik at the University of Utah and Myriad Genetics cloned the *BRCA1* gene in 1994. Mutations in the *BRCA1* gene can lead to breast cancer in 60% of women before 70 and can lead to ovarian cancer in 40% of women. Another gene linked to ovarian cancer and breast cancer is *BRCA2* on chromosome 13, cloned in 1995. The *BRCA2* gene mutations can lead to breast cancer in 45% of women before 70 and ovarian cancer in 25% of women. *BRCA1* and *BRCA2* are now known to be tumor suppressor genes that are particularly important in DNA repair. The *BRCA1* and *BRCA2* genes were patented by Myriad Genetics in 1998. However, after several court cases the US Supreme Court in 2013 ruled that naturally occurring genes could not be patented as they were products of nature. Rulings against the patents reduced the cost of testing for BRCA1/2. This ruling in the United States does not directly apply to other countries and there are subtleties around issues of manipulation of genes and diagnostic testing. Optimistically, a balance between biotechnology profits, research support, and the common good will prevail. These issues came to the fore with COVID-19.

In the last 20 years, so much sequencing information has been produced as the genomes of many microbes, plants, and animals were sequenced with improved sequencing techniques. This sequence information, together with PCR, makes gene discovery easier. However, assigning gene function to important processes and phenotypes can still provide challenges.

Paralleling advances in sequence information and PCR in gene discovery and function, particularly in human health, have been the bourgeoning use of DNA in fingerprinting for forensics, paternity testing, human ancestry analysis, species relationships, and ancient DNA. Finding a relatively simple unique fingerprint in humans is not so easy given most of the DNA between individuals is similar. In 1985, the same year as the publication of the PCR method, another revolutionary discovery was published by Alec Jeffreys. Jeffreys found that sizes of minisatellites, tandemly repeated sequences scattered across the genome, varied between individuals depending on the number of repeats in the minisatellite. Isolated DNA is cut with a restriction enzyme, run on a gel and denatured to separate the strands. A radioactive single-strand DNA probe for a given minisatellite is hybridized to the DNA on the gel to reveal bands of different sizes, depending on the number of repeats in the minisatellite. Jeffrey's technique almost immediately became prominent because of a complex immigration/paternity case and a rape/murder case in the United Kingdom. The principle for DNA fingerprinting and profiling was established, but the technique used to visualize the gel bands, though widely used in molecular biology labs (Southern blotting), was cumbersome. Several years later, the technique was refined by using PCR and simpler, shorter repeats known as short tandem repeats (STRs). New-generation sequencing advances mean that the genomic sequences of individuals can be readily obtained, enabling new approaches to disease diagnosis, evolution, and family lineages. The power of contemporary DNA technology to sequence and assemble genomes is illustrated by the work of Svante Pääbo, awarded the 2022 Nobel Prize for Physiology and Medicine. Pääbo, with many coworkers, obtained the DNA sequence of Neanderthals, our closest extinct relative who first appeared in the fossil record 400,000 years ago until 30,000 years ago. *Homo sapiens* first appeared in Africa about 200,000 years ago. The sequencing indicated that humans and Neanderthals coexisted and interbred, as did another DNA sequence from the hominin called Denisova. Both Neanderthals and Denisovans left traces of their DNA. This information helps further understand human physiology and ourselves.

Medical biotechnology has led to diagnostic tools that were not previously available as we have seen with the case of the *BRAC1* and *BRAC2* genes. This applies to other genes as well as virus and bacteria diagnostics. With the availability of rapid whole-genome sequencing, screening for less common deleterious genes is possible. Improved sequencing and analysis are also facilitating the understanding of the microbiome—our gut bacteria and how it influences health.

To move forward in other areas of medicine including stem cells and gene therapy, an important requirement is to have a vector that effectively and safely delivers genes into cells. While recent work on the discovery of disease-related genes due to deleterious mutation has accelerated, the ability to deliver nucleic acid cargo safely and readily for gene therapy has lagged. Recombinant adenoviruses are considered the first effective gene delivery vectors in humans. Human adenoviruses have been widely used and researched and were first used as vectors for mammalian gene therapy in the early 1990s. Adenoviruses are nonenveloped double-stranded DNA viruses, with a DNA genome of about 36,000 base pairs enclosed within its capsid. Intact adenoviruses cause respiratory tract symptoms of the upper respiratory track. There are over 50 serotypes that are useful in identifying the most effective vector. Replication-deficient recombinant adenovirus enter the nucleus, are not integrated into the genome, and there is no requirement for cell division. Following animal trials, there was good progress, with the first generation attenuated, replication-deficient adenovirus vectors developed (E1, E3 genes deleted in a cassette for gene insertion). It was shown that with the normal CFTR gene (mutation causes cystic fibrosis) therapeutic levels of mRNA could be produced in the human airways epithelium. However, in other human trials, there were difficulties with immune responses, including a trial fatality in 1999, with a gene directed to the liver. This sad event was understandably a setback for the field, but, 20 years later, gene therapy is undergoing a renaissance. There has been further modification of the vector (second generation) by removing additional genes, increasing the cloning capacity and reducing immunity. Adenovirus vectors provide expression of the transgene for up to 2 weeks. They are suitable for transferring genes to most organs and with current procedures are now regarded as safe, based on many clinical trials. Third-generation nonreplicative high-capacity adenovirus vectors (HCAdVs) or so-called gutless adenoviral vectors have been developed that are able to further reduce anti-adenoviral vector immunity due to the deletion of all viral structural genes.

Adenovirus vectors were originally developed for gene therapy where cells are infected with normal genes to overcome diseases caused by mutant defective genes. However, there are two other uses of adenoviruses. Adenoviruses can be used in vaccines where the expression for a period of a few weeks is fine. A well-known example is the use of an adenovirus vector for the COVID-19 AstraZeneca/Oxford DNA vaccine. In this case, a simian adenovirus (E1and E3 genes deleted) was used to further reduce preexisting vector immunity. Though mRNA vaccines were considered more effective than the AstraZeneca DNA vaccine, it was essentially effective and widely used (including my first two vaccinations). There were the clotting side effects in a small number (1 in 50,000 in people under 50) of people. The precise reason is unclear. Another potential use is in a therapy for cancer. In this case, the adenovirus is modified (oncolytic adenoviruses) to be conditionally replication competent and makes use of their lytic cycle to kill cancer cells. The oncolytic virus is modified to only replicate in cancer cells, recognized by having a mutation in a tumor suppressor gene.

Another vector now in use for a large number of clinical trials for monogenetic disease is the adeno-associated virus (AAV), which can be engineered to deliver DNA to target organs. AAV is not known to cause human disease. AAV is a nonenveloped single-stranded DNA virus of about 4 kb in length. Unusually, to replicate it requires the presence of another virus, usually an adenovirus. However, it can be used as a vector where replication is not required, persisting as an extrachromosomal element in the nucleus, not integrating into the host genome as do retroviruses. Of particular value is that different capsid proteins can be used to encapsulate the recombinant DNA. These different capsid proteins can target different tissues for therapy application and facilitate entry into the cell. An example (see Chap. 13) is the treatment of an inherited retinal disease due to a deleterious mutation in the *RPE65* (*RETINAL PIGMENT EPITHELIUM65*) gene. This disease causes severe vision loss. A working copy of the gene (marketed as Luxturna) is delivered to retinal cells. The first report of phase 3 trials in 2017 by Stephen Russell and coworkers showed improved functional vision. The gene therapy field is now advancing rapidly.

Another development that further increases the toolbox for gene therapy is the CRISPR/Cas9 gene editing tool, discussed more fully in Chap. 11. This has now been refined to modify specific bases using a base-editing tool or prime-editing tool. This means that a mutation in a disease gene can be corrected in a very direct way. There are now many tools that can advance gene therapy. There will no doubt be continued refinement of safety in and accuracy as clinical use is developed using vectors such as those discussed above. At this time, the future of gene therapy looks very bright.

The fullest application of biotechnology targeted to human health required the development of suitable vectors such as the adenovirus and AAV vectors. The fullest application of plant biotechnology also required vector development for gene delivery and is elaborated on in Chap. 11. Interestingly, it relied on the use of a plasmid in a bacterium (*Agrobacterium tumefaciens*), which causes tumors in plants. Key gene sequences were identified (and tumor genes deleted) that allowed gene insertion into plant genomes for increased plant yields, nutrition, and plant disease; and also for environmental benefit.

How is biotechnology being developed in animal agriculture and fish aquaculture? In cattle (as an example), there is *in vitro* fertilization initiated after the cloning of "Dolly" by somatic cell nuclear transfer (SCNT, Chap. 5). Cloning of cattle by SCNT has become part of breeding strategies. The types of genetic modification in cattle being used have different ethical and safety concerns compared to humans. Two strategies are used for gene editing in cattle. One is to transfer the construct by transfection (e.g., by electroporation) into a somatic cell line and then produce animals by SCNT. Alternatively, DNA is injected directly into the zygote. In both cases, subsequent embryo at the blastocyst stage is transferred to a surrogate to produce the fully developed offspring. One example is the demonstration that the production of poll cattle (hornless) by gene editing is possible. In aquaculture, there has been approval by the US FDA for the commercialization of GM Atlantic salmon, marketed as AquaAdvantage salmon. Eggs are genetically modified with a growth hormone gene, causing faster growth. The salmon is not released into the wild but grown and contained in tanks in inland areas.

Ethical use of animals has been critical for advances in clinical application. The COVID-19 situation again provides a good example as particular animals (mice, hamsters, ferrets, and nonhuman primates) were important in trials prior to clinical tests. Since 1953, there have been many key advances in DNA molecular biology and biotechnology to provide the potential that exists today, as well as contributing to a better understanding of ourselves. It is always difficult to read the future, but current knowledge already points to lots of possibilities. How genes link to the cellular landscape is ongoing. Two new areas that provide additional possibilities are biomolecular condensates and the details of protein structure. Increasingly it is apparent that groups of molecules are able to interact in structures that condense like droplets. Within these condensates, proteins drive specific reactions. It used to be a very big job to work out the structure of an important protein using such tools as X-ray crystallography and cryo-electron microscopy. Using artificial intelligence, it is possible to predict 3D protein structure. This artificial intelligence system is called AlphaFold, developed by Demis Hassabis and John Jumper. Further refinements are necessary to relate protein structure to various molecular interactions, but it is a major advance.

Further Reading

Berg P, Mertz JE. Personal reflections on the origins and emergence of recombinant DNA technology. Genetics. 2010;184:9–17.

Cohen SN, Chang AC, Boyer HW, Helling RB. Construction of biologically functional bacterial plasmids in vitro. Proc Natl Acad Sci USA. 1973;70:3240–4.

Crystal RG. Adenovirus: the first effective in vivo gene delivery vector. Hum Gene Ther. 2014;25:3–11.

Goeddel DV, Kleid DG, Bolivar F, Heyneker HL, Yansura DG, et al. Expression in *Escherichia coli* of chemically synthesized genes for human insulin. Proc Natl Acad Sci USA. 1979;76:106–10.

Mueller ML, Van Eenennaam AL. Synergistic power of genomic selection, assisted reproductive technologies, and gene editing to drive genetic improvement of cattle. CABI Agric Biosci. 2022;3:13.

Mullis KB. The unusual origin of the polymerase chain reaction. Sci Am. 1990;262:56–65.

Naso MF, Tomkowicz B, Perry WL III, Strohl WR. Adeno-associated virus (AAV) as a vector for gene therapy. BioDrugs. 2017;31:317–34.

Ricobaraza A, Gonzalez-Aparicio M, Mora-Jimenez L, Lumbreras S, Hernandez-Alcoceba R. High-capacity adenoviral vectors: expanding the scope of gene therapy. Int J Mol Sci. 2020;21:3643.

Yoshida K, Miki Y. Role of BRCA1 and BRCA2 as regulators of DNA repair, transcription, and cell cycle in response to DNA damage. Cancer Sci. 2004;95:866–71.

Zagorski N, Profile of Alec J. Jeffreys. Proc Natl Acad Sci USA. 2006;103:8918–20.

11

DNA and Food and Nutrition for 10 Billion People

The world's population is approaching 8 billion, and it is estimated that 10–11 billion may be reached by the end of the century, though when and at what number the population will peak is debated. It is clearly going to be a very big number. Plant genetic engineering (GE) has opened up many possibilities for plant improvement both in conjunction with and in addition to traditional plant breeding. This has become possible because of an elegant transformation method using the bacterium Agrobacterium tumefaciens *and its Ti plasmid. The plasmid is disarmed and advantage taken of the ability to insert DNA into the nuclear DNA of its host. Refined transformation systems are now available. For transformation, the gene is inserted into somatic cells in tissue culture, which are then regenerated to produce mature plants. Early biotechnology involved a foreign gene from a bacterium (*Bacillus thuringiensis*). This was the* Bt *gene for insect resistance. An important plant biotechnology success was the production of golden rice for vitamin A formation in 2000. This became the center point of public debate on GE plants. Other examples of biotechnology given are for virus resistance, gene editing for vitamin D enhancement, and using plant cells for vaccine production. Transformation has aided gene discovery, and the current emphasis is on using the understanding of the genetic regulation of plant processes to increase yield and respond to climate change.*

R. J. Rose, *DNA Know Thyself*,
https://doi.org/10.1007/978-981-97-8381-6_11

What is the food situation? In 2021, the United Nations' FAO estimated that 768 million people faced hunger and 237 billion people lacked nutritious and adequate food. Given the COVID-19 pandemic and war in Ukraine, invaded by its neighbor, things are worse. The effects of the pandemic have declined as vaccination reached across the world and antivirals became more available. Hopefully, major conflicts will be resolved. Suffice to say, we must bring all our knowledge to bear on sustainably providing enough nutritious food to feed humankind. In this chapter, I wanted to focus on plant agriculture as a primary source of our food, directly or indirectly.

In previous chapters, the nature of the DNA molecule, the gene, how it is regulated to produce a sophisticated multicellular organism like ourselves, and the rise of biotechnology have been considered. This understanding of genes and their regulation has helped explain how the environment is sustained and new approaches to disease treatment using advances in stem cell knowledge and gene therapy. What is happening to sustain adequate food and nutrition using DNA-related technology? With the staple food crops such as rice, wheat, and maize, genetics has been important in increasing productivity and in controlling disease. Application of genetics in plant breeding to produce high-yielding varieties was an important part of the green revolution. There were additional inputs such as fertilizer, irrigation, and pesticides. These latter inputs sometimes led to negative environmental impacts due to injudicious use and overuse. In some cases, these negative effects were due to policy settings (e.g., land use, subsidies) and were corrected, leading to greater sustainability. Clearly management practices need to be optimized, but there is a broader role for plant genetics and breeding for water and nutrient use efficiency, pest resistance, as well as yield increase, for food sustainability.

Advances in DNA biology and technology led to investigations into the transformation of plant cells to directly modify the genome to provide or enhance specific traits. The concept of DNA transformation was not new as it had long been part of bacterial genetics, dealing with the less complex prokaryote cell. The development of genetic engineering techniques in the eukaryote plants has an interesting history. Prior to the 1950s, it was known that a bacterium, *Agrobacterium tumefaciens,* caused a disease called "crown gall" in a number of species. This disease occurred at the base of the plant near soil level and was a mass of proliferating cells, a tumor, which choked off transport of nutrients in the plant stem. What was the bacterium doing? What was the tumor-inducing principle that caused the plant cells to proliferate? It wasn't until 1980 that the tumor-inducing principle was shown to be a piece of DNA (now known as T-DNA) transferred from the bacterium into the plant's DNA. Here then was a potential vehicle for transferring foreign DNA into a plant cell. It was subsequently found that what was key were the two border regions of the T-DNA and the tumor-inducing pieces (coding for the synthesis of the plant hormones auxin and cytokinin) could be removed. New genes could now be spliced between the two border regions with the necessary promoters to enable transcription. The Ti plasmid had another special piece of DNA known as the virulence or "vir" region, which enabled the T-DNA to enter the cell and be inserted into the plant's DNA.

To produce a transformed plant, another piece of biology was necessary; the ability to regenerate plants from cultured cells or tissue pieces. As indicated in Chap. 5, plant cells are totipotent and isolated cells can regenerate into a whole plant when cultured appropriately. By transferring the DNA to cultured cells, mature seed-bearing plants can be produced from the transformed cells. In 1983, the successful production of transgenic (genetically engineered) plants or plant cell lines was reported by three research groups (Chilton group, Schilperoort group, Schell/van Montagu group). At this point, the constructs for the *Agrobacterium*-mediated T-DNA transfer had not been optimized but this soon followed, so healthy plants containing new genes could be produced. The *Agrobacterium*-mediated DNA transformation has been greatly refined. Today, there are molecular "toolkits" that can be readily obtained to enable the transgenic plant with the gene of interest to be produced. The key remains the T-DNA border sequences and the virulence genes now housed in two separate plasmids. Plasmids are circular pieces of DNA that are separate from the genome (the bacterial nucleoid) but are also maintained by replication. The T-DNA plasmid can be precisely structured and inserted back into the *Agrobacterium* for DNA transfer to the plant cell. While other methods of plant transformation were developed, the *Agrobacterium* technique is the one that became the standard technique.

The development of the *Agrobacterium* transformation method had a dramatic effect on the improved understanding of plant genetics and the development of plant biotechnology, where genes could be inserted for crop improvement. To take advantage of transformation for a given species and a variety of that species, the plant needed to be regenerated after the DNA insertion. While principles have emerged, and the molecular genetics better understood, regeneration remains difficult in many cases. Nevertheless, it has been resolved for an increasing number of crops. A new technique for transferring genes with *Agrobacterium* has been discovered in the major model for plant functional genomics. The major model is a tiny plant called *Arabidopsis thaliana* (common name is thale cress). The floral dip transformation method, first published in 1998, involves dipping the inflorescence of the plant into an *Agrobacterium* solution. While this works well with *Arabidopsis,* it hasn't been successful in other species. Working with *Arabidopsis,* the SALK T-DNA mutant collection was produced in 2003 using *Agrobacterium*. The T-DNA is inserted randomly, so with a huge number of plants (at least 220,000) all of the approximately 30,000 genes have a DNA insert causing a mutation. This has enabled the identification of the function of many plant genes, revolutionizing plant genetics.

While the understanding of plant genes was advancing, plant biotechnology initially focused on inserting foreign genes for agricultural production purposes. A foreign gene directed against insect attack provides an example of this approach. There is a soil bacterium called *Bacillus thuringiensis* (*Bt*) that produces a protein toxin, coded by the *cry* gene, which is active against certain insects. *Bt* crops (i.e., containing the *cry* gene), including cotton, maize, and soybean, have been successfully commercialized since the mid-1990s. The CRY protein is not toxic to mammals. In susceptible insects, the toxin binds to cells lining the gut, causing a pore to form, which leads to insect death. The use of this gene would seem to be a win for agriculture and the environment, reducing the use of insecticides. There have also been refinements of management and gene technology to reduce resistance development. While many countries use *Bt* crops, it is not universal even with refinements of management and gene technology. The European Union, for example, does not allow the use of GM (genetically modified) crops. Why is this so? There are extensive requirements before the commercialization of GM crops in other countries. In the United States, the Environmental Protection Agency (EPA) found that there were no unreasonable risks to human health and the environment from

Bt crops. The banning of GM crops in some countries seems philosophical rather than scientific. It is argued that the use of foreign genes will have unintended consequences, despite extensive research and the safety record to date. This will be elaborated further in the context of gene editing, which is considered later in the chapter.

Another early GM approach comes from the development of genetically engineered (GE) virus-resistant crops. This is illustrated by the commercial success with papaya and its susceptibility to the *Papaya ring spot virus*, an RNA virus that is aphid transmitted. *Papaya ring spot* causes unsaleable fruit and eventually kills the plant. Excellent resistance can be developed using the introduction of the virus' coat protein gene. Coat protein-mediated resistance (CPMR) was first demonstrated in 1986 and in papaya in 1992. It seems likely that the CPMR mechanism in papaya acts at both the protein and RNA levels. The protein can influence virus uncoating and transport, while the RNA can suppress viral RNA synthesis due to double-stranded RNA interference (post-transcriptional gene silencing). There are no reports of negative environmental effects.

The difficulties in gaining acceptance of GMOs (genetically modified organisms) and negotiating commercialization, including regulation requirements, are seen in the case of "golden rice." This begins with the demonstration of how vitamin A levels in humans could be enhanced by introducing the vitamin A precursor β-carotene into rice grains. Rice in its milled form is characterized by the complete absence of vitamin A and its carotenoid precursors. In 2000, Ingo Potrykus and Peter Beyer reported the *Agrobacterium*-mediated transformation of the entire β-carotene biosynthetic pathway into rice endosperm. When eaten, the β-carotene is converted into vitamin A. The Potrykus and Beyer publication was not just an important demonstration of the potential power of GMOs but was of significance in the fight against vitamin A deficiency. Vitamin A deficiency can cause frequent infections, blindness, and even death. It has been estimated that vitamin A deficiency occurs in one in three children under the age of 5 and causes 670,000 deaths every year. It was thought that this would be valuable for children in Southeast Asia and in Africa where the main diet is rice. It is only now that this dream is being realized for a product for the public good.

The delay has been a frustration for Potrykus who has been very critical of the delay in authorization of golden rice use, arguing that it has cost lives. He considers GMO regulation is based on an "extreme interpretation of the precautionary principle." There are also patent problems in the development of GE crops, but as a "humanitarian project" this problem was solved with the help of an intellectual property expert (Adrian Dubock). In 2010, Potrykus outlined the specifics of how GE regulation had delayed the fruition of golden rice utilization by 10 years. There was then a very long process (including regulatory hurdles) of transferring the trait to a cultivar with yield equal to current non-GE crops. The cultivar to be released is called dhan 29 (suitable for the Philippines and Bangladesh), which has similar properties to the same cultivar without the β-carotene genes and was produced by the International Rice Research Institute (IRRI) in the Philippines. The IRRI project was licensed by Potrykus and Beyer as a nonprofit, humanitarian endeavor. Dhan 29 was produced by crossing with the first golden rice cultivar GRG2. This can be replicated with other cultivars optimized for a particular agricultural region. The Philippines approved the commercial propagation of golden rice and have started seed production. Regulators in the United States, Canada, Australia, and New Zealand have approved golden rice for consumption, illustrating a broadening acceptance of golden rice by regulators. A particular objective of the IRRI was to have the crop produced in Bangladesh, where about 21% of children have vitamin A deficiency. Approval is said to be imminent, after extensive field trials. The experience of rice biofortification for enhancing vitamin A, despite all its difficulties, has pioneered biofortification for human health. There is now a real possibility of bringing improved health to populations where there are nutrient deficiencies.

The examples of genetic engineering that have been considered involved genes from a bacterium, from a virus, and in the other cases from different plants. A new technique has emerged that has provided new approaches and has also altered the GMO "debate." This is gene-editing technology using CRISPR/Cas9, which was discovered by Emmanuelle Charpentier and Jennifer Doudna, who published a key paper in 2012. They were awarded the Nobel Prize for Chemistry in 2022. Gene editing allows DNA sequence changes at specifically targeted sites in the genome.

This ground-breaking technique has its origins in an ancient immune system that occurs in bacteria to eliminate invading bacterial viruses by cleaving their DNA using CRISPR/Cas9. CRISPR stands for *clustered regularly interspaced short palindromic repeats*, which have pieces of viral DNA between the repeats that get incorporated into the bacterial DNA. Cas stands for *CRISPR*-associated, the cutting or cleaving enzyme. Charpentier studied this primitive bacterial immune system in the pathogenic bacterium *Streptococcus pyogenes.* In her studies, Charpentier identified an RNA called tracRNA that is part of the CRISPR system, connecting CRISPR-RNA synthesized from CRISPR-DNA to Cas9. Doudna studied Cas proteins and small RNA molecules before collaborating with Charpentier. The viral *DNA* sequences between the CRISPR repeats serve as a memory of previous infections. With a new viral infection, CRISPR-RNA is formed by transcribing the CRISPR-DNA containing the viral sequences. The tracRNA recognizes sequences on the CRISPR-RNA and recruits Cas9 to form a complex at the viral sequences and Cas9 cleaves the virus recognition sequences, stopping the infection.

With the knowledge outlined above, Charpentier and Doudna collaborated to create the CRISPR/Cas9 gene-editing tool that can also work in eukaryotes. A single guide RNA (sgRNA) consists of the CRISPR-RNA (crRNA), which is the specific RNA sequence that targets a specific DNA sequence, plus tracRNA, which recruits the Cas9 endonuclease, to cut the double-stranded DNA at the specific site. The sgRNA and Cas9 are delivered to the cells separately. Once the DNA is cut, there are two main options. First, the cell's own DNA repair mechanism triggered by the double-stranded break is used. The endogenous repair enzymes that insert the joining nucleotides are error prone and can effectively mutate a gene of interest, important in gene function studies. Second, a new homologous DNA template can be inserted to precisely edit the genetic code of the gene of interest or modify promoter sequences, for example. Gene editing has continued to evolve to ensure accurate edits and prevent off-targets. Furthermore, there have been refinements of the CRISPR/Cas9 system that enable specific bases to be modified by the base-editing or prime-editing tools.

In 2022, a group led by Cathie Martin at the John Innes Centre in the United Kingdom employed gene editing to enhance vitamin D3 levels in tomatoes. Vitamin D can be synthesized by humans from 7-dehydrocholesterol (7-HDC), also known as vitamin D3, using ultraviolet B (UVB) in sunlight. Vitamin D deficiency affects skeletal development and has been implicated in increased risk of major diseases, including cancer, dementia, and the severity of COVID-19. The major source of vitamin D3 is dietary, particularly in fish and dairy. Approximately 1 billion people suffer from vitamin D deficiency. A plant source of vitamin D3 would be valuable but vitamin D3 levels in plants are low. The Martin group was able to increase vitamin D3 levels by knocking out a specific form of a competing 7-dehydrocholesterol-reductase (Sl7-DH2) using CRISPR-Cas9. This strategy was possible as there was a duplicated section of a relevant pathway, so knocking out Sl7-DH2 had minimal effects on synthesis of other pathways but enabled increased vitamin D3 levels.

Gene editing has widened the scope for acceptance of GMOs. As an example of the middle ground, Australia requires no regulation where no new genetic information is introduced, but there is a mutation. This is the case with CRISPR-Cas9 where DNA is cut at a specific site and the host cell repairs the break naturally, which may cause mutations to be introduced. However, where a new template is added at the cut site this is considered a GMO and is subject to GMO regulation. There are more lenient gene-editing rules in the United States, Brazil, and Argentina, and tougher measures in the European Union. China has approved gene-edited crops that knock out genes or have single-nucleotide changes, otherwise they are treated as GMOs with stringent regulation.

There are other interesting developments in GMO-related technology. One interesting example is in using plants for vaccine production. Virus-like particles (VLPs) of suitable proteins have been produced in plant leaves that are suitable for vaccine purposes. The VLPs are produced in the large leaves of *Nicotiana benthamiana* by transient expression. They are not regulated as GMOs, no DNA is inherited, and no nucleic acids are present in the VLP product. Leaves, acting as factories, are infiltrated with a solution of *Agrobacterium* with the appropriate gene constructs. The VLPs are then extracted for vaccine production. The biotechnology company Medicago Inc. has produced VLPs of the SARS-CoV-2 virus spike protein for a vaccine against COVID-19 called COVIFENZ®. This vaccine, approved by Health Canada, has been reported to be 71% effective in protecting trial participants against COVID-19.

Two examples of nutritional enhancement have been considered, and there is much potential in the nutrition area. It should be possible to regulate the storage products of seed so that the relative quantities of carbohydrate, dietary fats, and protein can be modified as necessary. More specifically the type of fatty acid and protein can be targeted. There has been great progress in understanding gene function and gene regulation in relation to growth and development, abiotic stress, and disease resistance. However, increases in plant yield using biotechnology have predominantly come from the use of the foreign *Bt* gene and herbicide resistance. The time has come to use the increased knowledge of plant genes and the regulation of plant processes. Understanding key yield determinants and the best genes to enhance is progressing (also see Chap. 12). In addition to pest and disease resistance, genes related to abiotic stressors such as drought, salt, high and low temperatures are being identified. Defining the best genetic strategies for resilience against climate change for a particular region, crop types and cultivars are emerging. What is needed is a refined pathway for identifying key genes, integrating these genes into key varieties and yield and resilience testing in the field. This would be more akin to what happens in medicine with moving discovery through to the different phases of clinical trials.

While GE biotechnology should have regulatory requirements, it is important that it is evidence-based. To date, GE has been safely integrated into plant breeding systems in a number of countries, but its full potential is still to be realized. In plant agriculture, it of course needs to be integrated with ecologically best practice. Genetic engineering in plants has the power to improve the well-being of many people and provide a new green revolution to feed the projected population.

Further Reading

Bhar A. Is it possible to ensure COVID19 vaccine supply by using plants? Nucleus. 2021;64:137–41.

FAO of the United Nations. The state of food security and nutrition in the world. Transforming food systems for food security, improved nutrition and affordable healthy diets for all. Rome; 2021.

Fernholm A. Genetic scissors: a tool for rewriting the code of life. Stockholm, Sweden: The Royal Swedish Academy of Sciences; 2020.

Jinek M, Chylinski K, Fonfara I, Hauer M, Doudna JA, Charpentier E. A programmable dual-RNA–guided DNA endonuclease in adaptive bacterial immunity. Science. 2012;337:816–21.

Li J, Scarano A, Gonzalez NM, D'Orso F, Yue Y, Nemeth K, et al. Biofortified tomatoes provide a new route to vitamin D sufficiency. Nat Plants. 2022;8:611–6.

Lindbo JA, Falk BW. The impact of "coat protein-mediated virus resistance" in applied plant pathology and basic research. Phytopathology. 2017;107:624–34.

Mendelsohn M, Kough J, Vaituzis Z, Matthews K. Are Bt crops safe? Nat Biotechnol. 2003;21:1003–9.

Potrykus I. Lessons from the 'humanitarian Golden Rice' project: regulation prevents development of public good genetically engineered crop products. New Biotechnol. 2010;27:466–72.
Somssich M. A short history of plant transformation. PeerJ Preprints. 2019;7:e27556v1.
Ye X, Al-Babili S, Kloti A, Zhang J, Lucca P, Beyer P, Potrykus I. Engineering the provitamin A (β-carotene) biosynthetic pathway into (carotenoid-free) rice endosperm. Science. 2000;287:303–5.

12

DNA and the Environment

In the previous chapter (Chap. 11), the potential for genetically modified (transgenic) plants to help feed the planet and provide nutrition was discussed. But how can DNA biology and technology contribute to a sustainable environment more generally? The themes I wanted to discuss are based on the importance of advances in DNA sequencing and analysis in understanding the biosphere and the progress in engineering nitrogen fixation and carbon capture by plants. The first part of the chapter examines the understanding of the soil microbiome that has come from new-generation DNA sequencing techniques. The soil ecosystem drives the plant ecosystem it supports. Similarly in the oceans, the microbiome drives the marine ecosystem. Nitrogen as part of our proteins and DNA is essential for life. Excess nitrogen fertilization and the energy required for nitrogen fertilizer production have become an environmental problem. Work is progressing in engineering nitrogen fixation in major crop plants, based on the understanding of the genetics of the rhizobium–legume symbiosis. Plants have an important role in combating climate change given their capacity for CO_2 fixation and their role in maintaining atmospheric homeostasis. Understanding the mechanisms of CO_2 fixation is enabling strategies to genetically enhance CO_2 capture. Reduced fertilizer use and reduced atmospheric carbon are now realistic propositions.

R. J. Rose, *DNA Know Thyself*,
https://doi.org/10.1007/978-981-97-8381-6_12

DNA sequencing and analysis is defining the biosphere in new detail. An organism's DNA signature helps in defining ecosystems and biodiversity and how this is being influenced by climate change. One aspect of ecosystems that has not been given emphasis is the soil environment. The upsurge in the interest and understanding of the gut microbiome has a parallel in the soil microbiome. The soil houses a complex ecosystem of its own, dependent on soil quality and type, the geographical location, and the aboveground ecosystem of plants and animals it supports. There are some interesting estimates indicating the biological significance of the soil microbiome. Soils may contain more than 1000 kg of microbial biomass of carbon per hectare, and 1 g of soil can contain around 10^8 microorganisms. DNA sequence analysis has opened up new approaches to unravel the soil's ecosystem from DNA samples, without culturing the microbes. Microbial culture is possible for only very small numbers. The soil microbiome in its entirety is being revealed for the first time.

DNA can be isolated directly from a soil sample using commercially available isolation kits. The initial isolation solution contains soil particles and different microbes. The microbial cells are lysed to release DNA from the cells, and the soil particles are then removed by centrifugation. The DNA-containing solution is then purified so that it can be used for amplicon sequencing or metagenomics using high-throughput shotgun next-generation DNA sequencing. Amplicon sequencing uses a standardized short sequence of DNA (a barcode) whose sequence of nucleotides is characteristic of the species. Examples of barcodes are 16s ribosomal RNA (rRNA) for bacteria and Archaea and 18 s ribosomal RNA for eukaryotes (soils include fungi and protists). Ribosomal RNAs are part of the ribosome where proteins are synthesized from the messenger RNA code. With metagenomics, most genes are sequenced giving additional information about the species.

The microbiome consists predominantly of bacteria and fungi, then Archaea and a small proportion of protists (predominantly unicellular eukaryotes). Quantification of a particular group can be determined by quantitative PCR. Unboxing of the black box of the soil microbiome can now be progressed.

The microbiome is in the bulk soil plus the soil in the vicinity of the plant roots (the rhizosphere). The dynamics of a given microbiome are related to the organic matter content, soil pH and moisture content, and the varied predators that feed on the bacteria and fungi, for example, protists, nematodes. Soil contains large amounts of carbon (C); globally more than the C in the atmosphere and vegetation combined. The soil carbon dynamic becomes an important part of C calculations in relation to climate change, particularly in relation to carbon sequestration. Microbial growth and activity are key in this regard. Microbes will release carbon as part of their catabolism, but also sequester carbon as a result of generating products that can be sequestered. The stable carbon pool consists of partly decomposed plant material, microbial products, and entombed microbial necromass due to dead microbes. Measuring the dynamics of the balance between carbon released from metabolism and the contribution of microbes to the stable C pool is an important metric. Where then does molecular biology fit into this? Unfortunately, only a fraction of the soil microbiome has been cataloged at species level. Though DNA analysis reveals the extent of the species diversity, culture of the most important species should still be a goal. The assembly of identified species by DNA sequencing will accumulate hand in hand with transcriptomics, proteomics, and metabolomics of different microbial ecosystems (e.g., forest soils, grassland soils, cropping and pasture systems). Soil ecosystem diversity is an important area for further research, particularly as carbon-emitting industries look for different types of offsets, including carbon sequestration.

The microbiome is also important in providing plant nutrients, particularly nitrogen and phosphorus. The decomposition of plant residues and other organic matter (known as mineralization) results in the formation of ammonium (NH_4^+). The ammonium is derived from the nitrogen held in organic form, and quantities will depend on the origin of the organic matter. In addition, using the genomic approaches of bar coding and metagenomics a coupling between C and N cycles can be shown. For example, decomposing maize residues by C-feeding bacteria can result in increases in bacteria involved in the nitrogen cycle.

Optimum plant growth is dependent on the availability of nutrients, particularly N and P, related to the activity of the microbiome. Microbes that interact with the plant more directly are in the layer immediately adjacent to the plant roots and are influenced by plant secretions. As discussed in Chap. 3, particularly important are the N-fixing rhizobia and the arbuscular mycorrhizal (AM) symbiosis influence on P uptake. While these latter specific interactions have been well studied, the dynamics of the rhizosphere microbiome, as for the soil microbiome, is in its infancy, but is now being addressed using DNA technology approaches. There are huge numbers of microbes in the rhizosphere. It has been estimated that there are 10^{11} microbes per gram of root and more than 30,000 prokaryotic species.

An interesting approach to the biological significance of the rhizosphere by Berendsen and coworkers is to conceptually link the rhizosphere microbiome to the gut microbiome, where in both cases microbes occur in abundance and diversity. First, there is the importance for nutrient uptake that is facilitated by the microbiome. This encompasses the N and P with rhizobia and mycorrhiza plus organic matter processing. Microbes in the rhizosphere tend to outcompete pathogens and prime the plant's immune system for defense. The ability to distinguish between beneficial and pathogenic microbes is a continuing challenge. It should be emphasized that plant secretions contribute to the microenvironment, encouraging beneficial bacteria in a mutualistic relationship. A metagenomics study in soil-grown sugar beets was able to identify microbiome characteristics that contributed to disease suppression. It was the abundance of certain taxa (notably Proteobacteria and Firmacutes) rather than differences in the number of different taxa that correlated with disease suppression. While this is illustrative of microbiome research, there is still much to be done in teasing out which microbes to focus on. This type of research is pointing to the potential of biofertilizers, where beneficial bacteria can be added to the soil.

There are other microbiomes in the biosphere that can now be investigated in more detail using metagenomics, especially that of marine ecosystems. It is not just a question of increasing the understanding of the species diversity in the ocean microbiome but how it interacts with other life forms and how it is being impacted by humankind. Oceans stretch across more than 70% of the world's surface, averaging more than 3.5 km in depth, with surprisingly microscopic organisms accounting for about two-thirds of the ocean's biomass. Microscopic organisms in the ocean contribute almost as much photosynthesis as land plants, so are important contributors to carbon sequestration. Craig Venter, the genomics pioneer, first initiated metagenomic sequencing in the marine environment, investigating the microbiome in the Sargasso Sea near Bermuda in 2004, and the discovery of new species and new genes has continued. Researchers in the field believe the majority of ocean microorganisms remain uncharacterized. This remains a rich source of novel genes. In 2004, the Venter study identified 1200 previously uncharacterized genes. What is going to be important is how the ocean microbiome responds to climate change and the effect on the ocean food web it sustains.

As discussed in Chap. 3, while most plants obtain their nitrogen as nitrate and ammonium from the soil, there are a group of plants known as legumes that can use atmospheric nitrogen. By avoiding fertilizers, using atmospheric nitrogen is so much more cost- and environmentally effective. The question arises as to whether it is possible using contemporary biotechnology to enable a wider range of plants to use atmospheric nitrogen. How do you enable a nonlegume to generate nodules housing rhizobia bacteria that fix atmospheric nitrogen and then transport ammonia to the plant for its amino acids? The nodule cells contain structures known as symbiosomes, which are surrounded by the plant-formed symbiosome membrane and house bacteroids. The bacteroids are the nitrogen-fixing form of rhizobia, using their enzyme nitrogenase. The bacteroids in the symbiosome are optimized to fix atmospheric nitrogen, particularly catering for nitrogenase's requirement for low oxygen levels to prevent its inactivation. The oxygen carrying hemoglobin, leghemoglobin, contributes to this regulation. Producing a nodule with specialized bacteroid-containing cells in nonlegume plants is no easy task. However, there is progress as many genes regulating the process have been discovered. A central problem is coordinating the development of the nodule with the bacterial infection—integrating these two processes.

There is an understanding of the key facets involved in forming an N-fixing nodule. Nodule organogenesis and bacterial infection via the infection thread require recognition of a rhizobial molecule known as Nod Factor (NF) induced by plant root secretions. NF (which has been chemically characterized as a lipochitin oligosaccharide) is recognized by a receptor on the root hairs of epidermal cells that activates a signaling pathway, in which Ca^{2+} has an important role. The consequence of this signaling is the activation of a protein kinase that is calcium and calmodulin dependent and designated CCaMK, which in turn initiates a signal-transduction pathway for nodule morphogenesis in the inner root cortical cells. The ability of a kinase to transfer phosphate groups is a common way to initiate signaling. Downstream of CCaMK, there is a requirement for a cytokinin for nodule formation. Cytokinins are important plant hormones that activate genes for growth and development roles. Many of the genes on the NF-induced pathway of nodule morphogenesis have been identified, and when some of these genes are highly expressed by genetic manipulation, nodules result. How do the bacteria populate the nodule cells? This is a fascinating process. The root hair, recognized by the rhizobium after NF perception, curls around the bacterium, forming a pocket. From this pocket, an infection thread develops by a combination of the divisions of the bacterium along a cellular tube-like pathway. NF perception has a role in setting this up by again activating key genes and initiating the infection thread. While many features of the signaling pathways are known, integrating the pathways by identifying key connections is required.

The identification of many of the controlling genes involved in the development of N-fixing nodules provides a genetic engineering opportunity to develop these organs in cereals like rice and wheat. Even knowing many of the genes, this would seemingly require the addition of multiple genes that can be integrated. So, this type of genetic engineering appears difficult. However, it is known that most plants can develop a symbiosis in their roots with arbuscular mycorrhizal fungi discussed in Chap. 3. This symbiosis utilizes similar molecules and genes to the rhizobia, legume symbiosis. For example, the AM fungi use an NF-type molecule (Myc factor) to induce symbiotic genes like those required for N-fixing nodules. Both NF and Myc factors activate the protein kinase CCaMK. Furthermore, comparative genomic studies between legumes and nonlegumes suggest that only a small set of genes is required. There is then a basis for optimism that the N-fixing cereals might be able to be engineered by introducing specific genes.

Researchers have also been trying another way of introducing nitrogen fixation into nonlegumes. This approach goes straight to the nitrogenase enzyme that is responsible for the nitrogen fixation by the bacteroids in the legume nodules. In essence, the strategy involves transfer of the bacterial nitrogen-fixing (*nif*) genes to the eukaryotic plant cell. This would eliminate the need for any direct involvement of bacteria. One of the biggest difficulties is providing the low-oxygen environment that nitrogenase requires. However, nature appears to have solved this problem in the photosynthesizing and oxygen-producing cyanobacteria. The cyanobacteria have been able to separate spatially or temporally nitrogen fixation from oxygenic photosynthesis. There may conceivably be a way of doing this in the plant chloroplast, which has evolved from the cyanobacteria. Again, a difficulty, but an exciting possibility. An intriguing recent discovery by Coele et al. and reported in *Science* in 2024 has been the discovery of nitroplasts that carry out nitrogen fixation in the unicellular algae *Braarudosphaera bigelowii*. The finding is consistent with the nitroplast evolving from an endosymbiont cyanobacteria. The nitroplast is a membrane-bound compartment of a eukaryotic cell, another model for engineering nitrogen fixation in eukaryotes. What is apparent is that there is a knowledge foundation for both engineering N-fixing nodulation and the introduction of nitrogenase. It is conceivable that with advances in biotechnology the introduction of nitrogen fixation into cereals will be a realistic proposition.

Another area open to advances in biotechnology is facilitating more carbon capture by the plant chloroplast, the organelle responsible for photosynthesis. Within the chloroplast is the enzyme ultimately responsible for life on earth, commonly known as Rubisco (ribulose 1,5 bisphosphate carboxylase/oxygenase). Rubisco is the most abundant enzyme in the living world. The Rubisco enzyme can capture CO_2 by adding one carbon (from CO_2) to a 5-carbon compound (ribulose 1,5- bisphosphate- RuBP) to form two 3 carbon compounds, and is responsible for most of the organic carbon in the biosphere. This then is the enzyme that is able to contribute to two of the biggest problems facing humanity—food production and carbon sequestration to ameliorate atmospheric carbon. Would it be possible to enhance the CO_2 fixation by improving the efficiency of Rubisco? Surprisingly, Rubisco turns out to be an inefficient enzyme for CO_2 fixation. Why is this so?

To understand why Rubisco is not efficient in fixing $CO_{2,}$ it is instructive to consider two groups of plants. Most plants, including wheat, rice, and soybean, are C3 plants. The first formed molecules as a result of CO_2 fixation by Rubisco are 3 carbon compounds. In C3 plants, two aspects of Rubisco are revealed. Not only does the enzyme react with CO_2 to capture carbon but can also react with oxygen (O_2) and RuBP to produce molecules with two carbons that enter into a process known as photorespiration. There is then a competition between CO_2 and O_2. Rubisco may have this dual capacity because of its evolution in a changing gaseous environment. Why not then engineer the Rubisco enzyme to block out O_2? It is interesting to look at what nature has done here to stop the oxygen effect.

There are another group of plants known as C4 plants, maize and sugar cane are examples. They are called C4 as an enzyme called PEP carboxylase captures CO_2 by adding carbon to a 3-carbon compound to form a 4-carbon compound. This 4-carbon compound is transported to another cell (bundle sheath cells) where Rubisco is present. CO_2 is released from this 4-carbon compound and concentrates around Rubisco crowding out O_2, increasing the efficiency of Rubisco for carbon capture. Another way of looking at C4 plants is that they have a CO_2 "pump," pumping CO_2 to Rubisco. Clearly Rubisco doesn't have to bind O_2 and this oxygenation doesn't have to occur.

Given that oxygenation of Rubisco can be prevented by concentrating CO_2 around it, there should be a way of modifying Rubisco to exclusively capture CO_2. Studies initiated to improve Rubisco have shown some improvement but are a work in progress. The approach that is being used is directed evolution (also see Chap. 14), which mimics natural evolution. A library of genes that translate into gene products that are functional variants are produced and tested. Strategies to maximize likely positive variants have been developed. In order to further understand the testing of Rubisco, it is necessary to understand the structure and genetics of Rubisco. Again, strange for such an important enzyme, there are two types of subunits: large subunit (LS) and small subunit (SS). The LS is coded for by chloroplast DNA and SS by nuclear DNA. They combine together in the chloroplast. The LS has been the main focus as it contains the active site where the CO_2 is fixed. To test for CO_2 fixation, the gene has to be inserted into chloroplast DNA, which is possible using biolistics with DNA-coated tungsten particles. Some researchers have also been experimenting with the SS based on evidence that it can influence the folding of Rubisco, indirectly affecting the active site on the LS. Both LS and SS research is augmented by Rubisco gene sequences in cyanobacteria and algae providing a diversity of inputs into Rubisco structure and function.

Another approach to improve carbon capture efficiency in C3 plants is to mimic the CO_2 "pump" in C4 plants by understanding the genetic changes associated with the evolution of C4 plants. A major difference between C3 and C4 plants is that in C4 plants the initial CO_2 capture occurs in mesophyll cells as the first step of concentrating CO_2 for the bundle sheath cells surrounding the vascular tissue where Rubisco fixes the CO_2 for photosynthesis. In C3 plants, the Rubisco fixes CO_2 in both mesophyll and bundle sheath cells. However, as we have seen for nitrogen fixation it is a question of focusing on the evolutionary genetic steps, in this case C4 from the more ancient C3 plants. Defining the core suite of genes, and their regulation, required for C4 photosynthesis using high-throughput sequencing of closely related C3 and C4 species is being pursued. The C4 rice project is supported by the Bill and Melinda Gates Foundation. C4 plant species are not the only organisms that can concentrate CO_2 for more effective CO_2 fixation. Cyanobacteria live in water and use CO_2 or HCO_3^- (bicarbonate) for photosynthesis. They can provide additional insights into genetic mechanisms for increasing CO_2 fixation in land plants. Rubisco is sequestered in a specialized structure known as a carboxysome and cyanobacteria have developed CO_2 concentrating mechanisms (CCMs) to deliver CO_2 to Rubisco.

Advances in DNA sequencing technology are enabling a more comprehensive understanding of the diversity of life on planet Earth. Biotechnology has opened up the possibility of dramatic advances in enhancing the utilization of atmospheric nitrogen and CO_2 capture by plants, with major improvements for agriculture and the environment. Reduced fertilizer use and reduced atmospheric carbon are now realistic propositions using current knowledge of genomics and the associated biological processes.

Further Reading

Abreu A, Bourgois E, Gristwood A, Troublé R, Acinas S, et al. Priorities for ocean microbiome research. Nat Microbiol. 2022;7:937–47.

Berendsen RL, Pieterse CM, Bakker PA. The rhizosphere microbiome and plant health. Trends Plant Sci. 2012;17:478–86.

Coale TH, Loconte V, Turk-Kubo KA, Vanslembrouck B, Mak WK, et al. Nitrogen-fixing organelle in a marine alga. Science. 2024;384:217–22. (Also see Perspective on this paper—Massana R. The nitroplast: A nitrogen-fixing organelle. Science. 2024, 384:160-1.)

Dubey A, Malla MA, Khan F, Chowdhary K, Yadav S, et al. Soil microbiome: a key player for conservation of soil health under changing climate. Biodivers Conserv. 2019;28:2405–29.

Mishra A, Singh L, Singh D. Unboxing the black box—one step forward to understand the soil microbiome: a systematic review. Microb Ecol. 2023;85:669–83.

Nannipieri P, Penton CR, Purahong W, Schloter M, Van Elsas JD. Recommendations for soil microbiome analyses. Biol Fertil Soils. 2019;55:765–6.

Oldroyd GE, Dixon R. Biotechnological solutions to the nitrogen problem. Curr Opin Biotechnol. 2014;26:19–24.

Suzaki T, Kawaguchi M. Root nodulation: a developmental program involving cell fate conversion triggered by symbiotic bacterial infection. Curr Opin Plant Biol. 2014;21:16–22.

von Caemmerer S, Quick WP, Furbank RT. The development of C4 rice: current progress and future challenges. Science. 2012;336:1671–2.

Wilson RH, Alonso H, Whitney SM. Evolving *Methanococcoides burtonii* archaeal Rubisco for improved photosynthesis and plant growth. Sci Rep. 2016;6:22284.

13

Gene Therapy for Human Disease

A major objective of gene therapy research and current treatments is to correct mutations that are responsible for disease. An improved understanding of DNA cell and molecular biology has led to the use of genetic engineering strategies and gene repair mechanisms. The chapter starts with engineering of T cells for CAR-T cell treatment of acute lymphoblastic leukemia. Isolated T cells are genetically modified to produce chimeric antigen receptors (CARs) for the treatment of leukemia. A gene addition strategy to overcome gene mutations is considered for two diseases. These are inherited blindness due to mutation of the RETINAL PIGMENT EPITHELIUM65 (RPE65) gene and hemophilia B due to deficiency or dysfunction of the Factor IX protein. Gene-editing approaches have been applied to a number of disease states. Homology-directed repair (HDR) and base editing to correct sickle cell disease are discussed. In addition to gene editing, epigenetic modification without sequence change, using a modified CRISPR-Cas, is another approach to modify gene expression. The final part of the chapter deals with the treatment of mitochondrial disease by mitochondrial replacement treatment (MRT) and a novel mitochondrial base-editing system. This latter topic raises the question of germline therapy.

A now-established gene therapy uses CAR-T cells to engineer the T cells of the immune system to treat blood

R. J. Rose, *DNA Know Thyself*,
https://doi.org/10.1007/978-981-97-8381-6_13

cancers. CAR-T refers to chimeric antigen receptor (CAR) T cells. T cells usually target foreign cells and virus-containing cells, and by a range of actions, including activation of other cells, can kill the targeted cancer cells. The T cells can be genetically modified to produce a specific surface protein (the CAR) that can recognize and kill specific cancer cells that are targeted by the introduced CAR. To engineer CAR-T cells, T cells are harvested from the peripheral blood and purified by cell separation techniques. The CAR is introduced by genetic engineering techniques, with the retroviral lentivirus used as a vector. The lentivirus vector has pathogenicity genes removed, can transfect the cell, and is designed for the insertion of the CAR DNA into the T cell genome using mRNA specifying the CAR. Being a retrovirus (see Chap. 9) vector, the RNA is reverse transcribed into DNA, and with the lentivirus vector the DNA can be inserted into the genome of both dividing and nondividing cells. The CAR consists of three domains: an ectodomain, a transmembrane spacer domain, and an intracellular endodomain. The extracellular ectodomain targets the cancer cell antigen and the endodomain is a signaling/activation domain that activates the T cell function. The CAR design has improved over several iterations by modifying the signal transduction domain to improve T cell activation of the immune system to destroy the cancer cells. Once the CAR-T cells are produced, their numbers are increased by cell culture before being transfused into the patient.

The US Food and Drug Administration (FDA) approved the first CAR-T therapy against acute lymphoblastic leukemia (a B cell lymphoma) in 2017. This approved CAR-T therapy from Novartis is known as Kymriah or Tisagenlecleucel and approval of other CAR-T therapies has continued. Cappell and

Kochenderfer, in a recent review of CAR-T outcomes, indicate that T cell therapy is now established as a treatment for "relapsed and/or refractory B cell lymphomas, B cell acute lymphoblastic leukemia and multiple myeloma." These hematological malignancies are treated with CAR-T cells targeting the CD19 cell surface receptor of malignant B cells. B cells are the antibody-producing cells of the immune system. These results are all very positive, but while it can be fully curative, maximizing remission, with reduced side effects, requires continual research. Cost in the order of $500,000 per patient in the United States is also an issue requiring research focus. However, an essentially similar treatment known as NexCAR19 produced by a company in India is reported to cost 10% of the price (Mallapaty R. 2024, News in focus, *Nature*. 627: 709–10).

Given the success with hematological malignancies, attention has turned to solid tumors. Finding specific tumor-specific antigens can be difficult, and the CAR-T cells have to traffic from the blood into the tumor. The tumor itself represents a hostile microenvironment for T cells. A recent advance has been the use of a strategy known as LINK CAR-T in which T cells require binding of two antigens with linked signaling, which reduces off-targets against normal cells. The use of CAR-T therapy against solid tumors remains a worthwhile challenge.

Many human diseases are caused by deleterious mutations, and correcting them in somatic cells (i.e., not germline) is now making good progress. There are two general strategies that have been used. The first strategy is to insert a normal gene to take on the function of the aberrant gene. This is essentially a gene addition strategy. The second strategy is to correct the mutation in the gene using the gene-editing approach involving CRISPR/Cas9.

There are successful examples of gene addition directed against diseases caused by a defective gene. Inherited blindness can be caused by a deleterious mutation in the *RETINAL PIGMENT EPITHELIUM65 (RPE65*) gene. A defective *RPE65* gene causes the photoreceptors to have an impaired response to light, and they will ultimately degenerate. The reversal of blindness can be obtained by delivering a normal copy of the *RPE65* gene. The vector carrying the gene is an adeno-associated virus (AAV2), which is delivered by a subretinal injection. The AAV (discussed in Chap. 9) is not a retrovirus, so it is not inserted into nuclear DNA, but exists as an extrachromosomal element in the nucleus. In 2017, the RPE treatment became the first FDA-approved gene therapy product for a genetic disease, marketed as Luxturna. Durable rescue of vision for 5 years has been reported.

Another example of gene therapy involving the delivery of a new gene is the case of hemophilia B, which is a sex-linked, inherited bleeding disorder due to a recessive gene on the X-chromosome. Hemophilia B leads to bleeding episodes in joints, muscles, and internal organs. It is a rare disease estimated to affect 1 in 40,000, predominantly men. The cause of hemophilia B is the deficiency or dysfunction of a protein called Factor 1X. This plasma protein has an essential role in blood clotting. The transfer of a functional gene using an AAV viral vector (AAV8) can return Factor IX to near-normal levels and reduce or prevent bleeding episodes. As indicated for *RPE65* gene therapy, the *FACTOR IX* gene is not incorporated into the genome but exists as an extrachromosomal element, which is transcribed to produce the protein. The AAV8 vector is administered as a single administration by intravenous infusion and is expressed in the liver.

The FDA approved the AAV-based gene therapy for hemophilia B, known as Hemgenix (etranacogene dezaparvovec), in November 2022. According to a *Nature* editorial in April 2023, the cost of gene therapy remains a challenge. Gene therapy typically costs US$1 million for a single treatment while for Hemgenix the current cost is US$3.5 million. While the refining of treatments and gathering treatment statistics is important, protocols for cost reduction need to be developed.

Gene Therapy 2.0 arrived with gene editing, based on the discovery of the CRISPR-Cas9 system. The discovery of the CRISPR-Cas 9 system was considered in Chap. 11. An example of its use in Chap. 11 for plant genetic engineering was the knocking out of the function of a specific gene. The gene-editing approach also enables the correction of the mutation in the gene responsible for a disease. The CRISPR-Cas 9 system as originally described cuts DNA at specific sites and allows insertions or deletions (indels) or a DNA template specified by the investigator. Indels can disrupt the function of a gene (as in the Chap. 11 example). More refined editing has been made possible by the discovery of several proteins that are related to Cas9. These new editing tool kits can be directed to specific requirements. There are base editors that can edit certain single bases and prime editors that can install all possible single base conversions. Both latter editors do not use double-strand DNA breaks. In addition, there are Cas transposases that can insert large DNA sequences, greater than 1000 nucleotides. For gene editing, there is the requirement for suitable delivery systems. Isolated cells can be edited (e.g., hematopoietic stem cells) and then returned intravenously. Delivery of CRISPR-Cas9 into isolated cells can be by virus or by electroporation of nonviruses where transient tiny pores allow entry into the cell. Injection into tissues can be via virus packaging and lipid nanoparticles can be delivered intravenously.

Gene editing is now an approach for combating sickle cell disease (SCD). Sickle cell anemia is due to abnormalities in hemoglobin, the oxygen-carrying molecule present in red blood cells. Hemoglobin A (HbA) consists of four protein chains (a tetramer), two α chains and two β chains. These chains are organized as dimers designated as $\alpha_1\beta_1$ and $\alpha_2\beta_2$, which associate to form the tetramer. The mutated form of hemoglobin in sickle cell anemia is referred to as hemoglobin S. HbS is due to a mutation in the β chain of hemoglobin. A single base change from thymine to adenine occurs, causing the substitution of a valine amino acid for glutamic acid. This single amino acid change has dramatic consequences, causing interaction between hemoglobin molecules, leading to aggregation of hemoglobin molecules and the sickle appearance of the red blood cells. These sickle cells can result in anemia and capillary blockage.

Gene editing to correct the sickle mutation has proceeded to the stage of clinical trials after demonstrating editing effectiveness. Two types of editing approaches have been used to correct the glutamic acid to valine change in hematopoietic stem cell engineering. One approach is homology-directed repair (HDR), using CRISPR-Cas9 to cut the DNA at the required site and then deliver, via homologous recombination, a template with the correct gene sequence. To achieve the most effective editing, the HDR response requires the cells to be dividing, delivering HDR when cells are in the G2 and S phases of the cell cycle. The HDR approach requires three components; the two CRISPR-Cas9 components, that is, the single guide RNA (sg RNA) that includes the target sequence in the genome to be changed and a Cas9 recruiting sequence, the Cas9 nuclease cutting enzyme to cut the DNA sequence and the

HDR repair sequence. There are three different ways of packaging the sgRNA and Cas9 (the cargo) for delivery. First, as DNA coding sequences in a plasmid, second as mRNA Cas9 and sgRNA, and third as Cas 9 protein and sgRNA. Delivery of the cargo in all cases can be by electroporation, where the tiny pores generated allow entry of these large molecules. In the case of the DNA approach, a virus such as AAV can also be used. The DNA sequence for HDR can be delivered by electroporation or AAV.

Rather than the HDR approach, the newer base editing can be used. Base editing utilizes a modified Cas9 known as nCas9 that only cuts a single DNA strand. The sgRNA and nCas9 target and cut a single DNA strand and with a deaminase enzyme linked to nCas9 edit the base. In SCD, there is a change from glutamic acid to valine in the β chain of hemoglobin. Valine has different properties to glutamic acid causing the hemoglobin abnormalities. What is then required in the case of SCD is the direct conversion of a single base. A point mutation in the glutamic acid codon GAG occurs, causing the A to be changed to a T, making the codon GTG, which is valine. While it is not possible to revert valine to glutamic acid by adenosine base editors, adenosine base editors can convert A-T base pairs to G-C pairs forming the GCG codon that encodes alanine, which also produces a nonpathogenic hemoglobin without sickle cell occurring.

The examples of gene editing to overcome SCD dealt with manipulating isolated blood stem cells *in vitro* then injecting back into the body. There are many diseases involving gene mutations in specific body organs and tissues that present difficulties in delivering the cargo for effective editing. Progress for the two examples is considered below.

Transthyretin amyloidosis (ATTR amyloidosis) is a monogenic disease that is progressively fatal due to the accumulation of an aberrant transthyretin (TTR) protein that does not fold properly. The misfolded protein forms amyloid fibrils. TTR is produced almost entirely in the liver and is circulated in the blood plasma. ATTR can cause cardiomyopathy and heart failure and/or amyloid polyneuropathy, a disease of the peripheral nervous system.

The approach here was to reduce the concentration of circulating TTR by using CRISPR-Cas9. In this case, after the cutting of the DNA, there is nonhomologous end joining resulting in small indels. The indels meant that the gene sequence is now disrupted and functional TTR is no longer produced. The lowering of the TTR results in the amelioration of the disease. In the study discussed here, the CRISPR-Cas9 editing cargo is delivered by lipid nanoparticles containing the sgRNA and the mRNA encoding Cas9. The nanoparticle is delivered intravenously and in the circulation the lipid nanoparticles adsorb ApoE proteins on their lipid surface. These APOE proteins bind the lipid nanoparticles to the liver cells for cargo delivery by endocytosis. The sgRNA targets the TTR gene for Cas9 cutting, and the indel insertion at the cutting site prevents functional TTR synthesis.

A small phase 1 clinical trial using the above strategy was carried out with patients with hereditary TTR (hTTR) causing polyneuropathy, with or without cardiomyopathy. At 28 days after infusion, one group of three patients showed a 52% reduction in TTR protein (0.1 mg/kg dose) and an 87% reduction with a 0.3 mg/kg dose, in another group of three patients. There were few adverse effects.

Another disease requiring gene delivery to the body and is also amenable to editing strategies is Duchenne muscular dystrophy (DMD) disease. DMD is a muscle-wasting disease caused by DNA sequence changes that disrupt normal production of the dystrophin gene. Dystrophin is a scaffolding protein that maintains the integrity of striated muscle. The disease affects 1 in 5000 boys. Boys are more affected as they have only a single X chromosome and it is a recessive X-linked disease. DMD is a very serious disease that leads to mobility problems, skeletal deformity, respiratory insufficiency, and cardiomyopathy, resulting in death by age 30. There are thousands of dystrophin mutations, mainly exon deletions (65–72%) but also point mutations (25–35%) and duplications. CRISPR-Cas9 has been used successfully in human cells and mouse models to correct dystrophin mutations and enable dystrophin synthesis. Editing can overcome exon deletions and point mutations. In the case of exon deletions, the exon can be replaced using CRISPR-Cas9, with HDR replacing the appropriate sequence. Base editing can be used for point mutations. The major problem is gene delivery to enable the editing that has been shown to overcome this monogenic disease. Gene delivery strategies are usually viral or lipid nanoparticles. The AAV vector has yielded positive results for DMD in mice and dogs, and research is advancing rapidly in humans. AAV has minimal immunogenicity, and there is long-term expression of the introduced DNA. A disadvantage is that the cargo capacity is approximately 4.7 kb, smaller than other viruses. However, it is possible to use a dual AAV system with cargo delivered in two vectors.

The examples given in considering CRISPR editing have involved the correction of gene sequences. However, in addition to DNA sequence, genes can be regulated by epigenetic modifications (we can call these marks) on promoters. These marks can be on the DNA (e.g., methylation) or on the histones that the DNA wraps around (e.g., methyl or acetyl groups). The epigenome has been discussed in Chap. 5, in relation to cell fate and the different cell types in a multicellular organism. In an important adaptation of

CRISPR-Cas9, the Cas can be modified in such a way that it doesn't cut sequences targeted by the guide RNAs. The modified Cas is known as dCas, with "d" standing for deactivation of the Cas cutting ability. The dCas can be linked to gene regulatory proteins that act as epigenetic effectors. These epigenetic modifications by the CRISPR-dCas system can up- or downregulate gene expression. For example, genes with a promoter that is switched off by methylation can be activated by removing methyl groups using CRISPR-dCas. The CRISPR-dCas system is also a remarkable technical advance for understanding the epigenome of different cell types and in stem cell biotechnology. By targeting a key gene (*PCSK-9*), epigenetic editing of the enhancer-promoter of this gene in mice has been used to reduce cholesterol levels.

An unusual form of gene therapy is the treatment of mitochondrial disease by mitochondrial replacement therapy (MRT). Mitochondrial disease is caused by mutation in mitochondrial DNA (mtDNA) and can be associated with different organs, including the heart and the brain. While not universally approved, it has been approved in the United Kingdom and Australia. MRT replaces the mitochondria with defective mtDNA with healthy mitochondria containing the correct mtDNA sequences. The MRT requires what is known as "three-person" embryos. To begin with, the egg that is donating the healthy mitochondria has to have its nucleus removed. The donor mitochondrial egg is then the recipient of the mother's nucleus, which contains all the genes except for the mitochondrial genes, which comes from the donor. This egg with the recipient mitochondria is then fertilized with male sperm. The MRT technique for mitochondria replacement is in its infancy, but can work, and leads to normal development. A report in 2023 indicated that in the United Kingdom children with replacement mitochondria have been born, but little information was available. A concern that had been raised is that in the nuclear transfer procedure there is at least some carryover of some of the defective mitochondria into the cell containing the healthy mitochondria. There is however a potential alternative to MRT, which is mitochondrial DNA editing. The CRISPR-Cas system cannot work in mitochondria as the guide RNAs cannot be delivered across the mitochondrial membrane. A novel base-editing system has been developed that does not use CRISPR. These base editors have been developed to modify mtDNA via T:A to C:G transitions. The CRISPR-free base editors are able to be targeted to mitochondria and directed to DNA sequences via transcription activator-like effectors (TALEs), which are nucleases that can be engineered to bind to desired DNA sequences.

Editing of mitochondria raises the question of heritable germline engineering. Editing the eggs or embryos would deliver normal mitochondria to the whole body, rather than the difficult job of correcting mutant mitochondria in complex organs such as the heart or the brain. In the case of mitochondria, MRT is available as an approved option, though more data collection from offspring is important.

The big questions of heritable germline gene engineering involving egg, sperm, or embryos concern the nuclear encoded genes where there are numerous disease variants. Germline engineering is widely used in animals and plants. However, there is a need to ensure DNA biology in human embryos is fully understood using *in vitro* systems. As discussed further in Chap. 15, it was reported at a conference in November 2018 by He Jiankui that he used edited human embryos to produce a live birth. This was condemned worldwide, and regulatory agencies worldwide agreed that such work, resulting in pregnancies, should not be carried out. Correcting gene function to prevent disease is one thing, but would it stop there? Providing safe and effective gene editing of somatic cells seems a positive and ethical use of this technology and is the best way forward.

Further Reading

Adashi EY, Rubenstein DS, Mossman JA, Schon EA, Cohen IG. Mitochondrial disease: replace or edit? Science. 2021;373:1200–1.

Cappell KM, Kochenderfer JN. Long-term outcomes following CAR T cell therapy: what we know so far. Nat Rev. Clin Oncol. 2023;20:359–71.

Doudna JA. The promise and challenge of therapeutic genome editing. Nature. 2020;578:229–36.

Fala L. Kymriah (tisagenlecleucel) for young patients with acute lymphoblastic leukemia: first FDA-approved gene therapy. 2018, The Oncology Pharmacist. Third annual guide to new FDA approvals—FDA updates, Leukemia.

Kan MJ, Doudna JA. Treatment of genetic diseases with CRISPR genome editing. JAMA. 2022;328:980–1.

Maguire AM, Bennett J, Aleman EM, Leroy BP, Aleman TS. Clinical perspective: treating RPE65-associated retinal dystrophy. Mol Ther. 2021;29:442–63.

Min YL, Bassel-Duby R, Olson EN. CRISPR correction of Duchenne muscular dystrophy. Annu Rev. Med. 2019;70:239–55.

Pipe SW, Leebeek FW, Recht M, Key NS, Castaman G, et al. Gene therapy with etranacogene dezaparvovec for hemophilia B. N Engl J Med. 2023;388:706–18.

Schuster SJ, Bishop MR, Tam CS, Waller EK, Borchman P, et al. Tisagenlecleucel in adult relapsed or refractory diffuse large B-cell lymphoma. N Engl J Med. 2019;380:45–56.

White SL, Hart K, Kohn DB. Diverse approaches to gene therapy of sickle cell disease. Annu Rev. Med. 2023;74:473–87. [Gene editing pp. 481–483]

14

Synthetic Biology: A New Frontier

Synthetic biology applies engineering principles to living cells and organisms. It builds on the understanding of DNA molecular biology (Chap. 4) and biotechnology (Chap. 10) in using genetic tools to engineer synthetic cellular systems to produce products and understanding. We can use an analogy of electrical circuits in the way cellular systems integrate their biochemical machinery. The chapter will illustrate, through examples, how systems can be designed to produce specific products or outcomes. Plant-derived artemisinin used in malaria treatment is difficult to consistently supply the quantities required by manufacturers, so synthetic biology is used in yeast (Saccharomyces cerevisiae). In the process, three major strategies enable a synthetic pathway for the production of artemisinic acid. Another engineered plant product is leghemoglobin used in laboratory foods. Synthetic biology can also be used to produce industrial products and microbes with diverse function. Directed evolution is an important tool in synthetic biology that can be used in improving the efficiency of enzymes as catalysts for a variety of purposes—industry, environment, and medicine. Another tool for protein engineering is the use of unnatural amino acids that enable a greater range of properties for a protein. As synthetic biology advances, more complex gene circuits that can be programmed into the cell are being

R. J. Rose, *DNA Know Thyself*,
https://doi.org/10.1007/978-981-97-8381-6_14

developed. An example is MultiFate, which uses artificial transcription factors.

In living systems, there are many biochemical pathways that sustain life, and depending on the environment the organism lives in, can produce all sorts of molecules. One way of looking at these systems is to think of electrical circuits as the way cellular systems are integrated. Potentially these circuits can be modified using genetic tools to build pathways with potential in manufacturing, agriculture, human health, and the environment.

Artemisinin, an antimalaria drug, was discovered in 1972. The major contributor to the discovery was the Chinese scientist Tu Youyou. Professor Tu received the Nobel Prize in Physiology and Medicine in 2015. Artemisinin was discovered in an extract from the plant *Artemisia annua* (sweet wormwood), used in traditional Chinese medicine. Artemisinin acts well against the most virulent causative parasite of malaria, *Plasmodium falciparum*. Plant-derived artemisinin is difficult to consistently supply the quantities required by manufacturers, so synthetic biology is used in yeast (*Saccharomyces cerevisiae*). Today, artemisinin is used in artemisinin combination therapy (ACT) in combination with another effective antimalarial drug to reduce resistance.

Yeast is engineered to produce artemisinic acid. Artemisinic acid is then chemically converted to artemisinin. The engineering of yeast to produce artemisinic acid is an example of the power to engineer metabolism to produce an important medical drug. Three major strategies enable a synthetic pathway to produce artemisinic acid. The first part of the pathway involves increasing the expression of the genes present in yeast that encode the molecules for the mevalonate pathway to produce a compound known as farnesyl pyrophosphate (FPP). The next part is the insertion of the amorphadiene synthase gene (*ADS*) to produce amorphadiene from the FPP product of the mevalonate pathway. Another enzyme, CYP71AV1, is introduced to facilitate the conversion of amorphadiene to artemisinic acid in a three-step reaction. The key genes *ADS* and *CYP71AV1* are obtained from *A. annua* and transformed into yeast. Chemical synthesis is used to convert artemisinic acid to artemisinin. Molecular tools in setting up the artemisinic acid pathway are the ability to upregulate genes with powerful promoters, suitable foreign enzymes, and the ability to reliably transform yeast. It should be noted that work continues on the *A. annua* plant in order to increase the yield of artemisinin in the plant system.

The type of metabolic engineering used in the production of artemisinic acid can be applied to the production of many products using *S. cerevisiae* or *E. coli.* The main type of manipulations required are the increased expression of genes encoding relevant enzymes in key existing pathways, blocking any competing metabolic pathways and introducing foreign enzymes necessary for the production of the required final product. Promoter strength is an important aspect of controlling rates of gene expression. CRISPR-Cas9-related approaches can be used to place genes in an appropriate position on the chromosome.

Another interesting plant product is leghemoglobin, which is used in plant-based, simulated beef burgers. This is part of the revolution in the production of "laboratory" foods. Regular beef burgers contain small amounts of blood, with the red color due to hemoglobin. The red color of hemoglobin is due to red heme iron (organic compound surrounding a central iron atom) associated with the globin proteins in the hemoglobin. The hemoglobin contributes to the flavor, nutrition, and aroma of the beef burger. Leghemoglobin is found in the nodules of legumes such as soybeans where it contributes to the low oxygen levels required for the action of the nitrogenase enzyme required for nitrogen fixation (see Chap. 12). Leghemoglobin is a red color, like hemoglobin, due to the presence of heme iron that gives the pinkish color to an active nodule. Isolation of leghemoglobin from nodules is not economically feasible, so a yeast strain has been engineered for its production.

The yeast *Pichia pastoris* (rather than *S. cerevisiae*), which has a high capacity for protein secretion, is used to produce leghemoglobin. Leghemoglobin consists of heme iron and a single globin polypeptide chain. To produce leghemoglobin requires high rates of synthesis of both the heme and globin protein components. All the native heme biosynthesis genes are overexpressed using a strong promoter and the foreign soy leghemoglobin protein gene (i.e., the legume globin gene) is introduced in multiple copies. The engineering is carried out using standard transformation and also CRISPR-Cas9 to integrate genes into the appropriate location on the chromosome. Similar to the production of artemisinin, there is a combination of enhanced native pathways and introduced key genes required for the synthesis of the product.

In addition to plant natural products, production of many diverse products is being pursued using metabolic engineering. Higher alcohols such as isobutanol, which are superior biofuels compared to ethanol, are being increasingly researched. The advantages of isobutanol include higher energy density and higher octane number (a measure of fuel performance). Biosynthetic approaches in *E. coli* involve increased expression of genes involved in the synthesis of 2-keto acids and the addition of two enzymes from other organisms for the conversion to isobutanol. Illustrating how broad synthetic biology has become is that synthetic approaches are being used for the development of new materials such as biofilms and cellulose-based materials.

The examples of synthetic biology considered so far have involved using bacteria and yeast to produce products via metabolic engineering. However, microbes themselves can be the final product after modifications using synthetic biology. A recent example has been the use of a bacterium to provide nitrogen as a biological fertilizer. In previous chapters (Chaps. 3 and 12), the importance of nitrogen for sustainable agriculture and the environment was emphasized. In quite a neat idea, the bacterium *Klebsiella variicola* strain 137 (Kv137) is utilized to provide a biological source of ammonia, taking advantage of the nitrogen fixation of Kv137. Kv137 is found in the rhizosphere of roots and on the root surface. The work described here was carried out with corn and was tested in the US corn belt. The idea was to have the nitrogenase gene constitutively active (on at all times) irrespective of the nitrogen levels in the soil. What was necessary was to remodel the control of the nitrogenase gene. The nitrogenase gene is under the control of two genes *nifL* and *nifA,* which are responsible for turning off nitrogenase when nitrogen levels are low, so nitrogen fixation is off when it is most needed for supplying nitrogen to the plant. To maintain continuous ammonia production, the genome was modified to delete the *nifL* gene and replace it with a constitutive promoter to produce Kv137–1036, which allowed the continuous expression of the *nifA* gene and continuous nitrogen fixation by the nitrogenase gene. Field trials showed increased yields when applied as a liquid suspension at planting. Toxicity and pathogenicity studies indicate that Kv137–1036 is a *K. variicola* strain that does not show toxicity or human infection. The kv137-1036 has been commercialized as Pivot Bio

PROVEN. How widely used this approach will become will require more experience in its field use.

The engineering of the soil microbiome offers opportunities beyond nitrogen. It is possible, for example, to generate microbes with phosphate-solubilizing ability to increase the availability of soil phosphate to plants. Further opportunities will be opened up as the soil microbiome becomes more completely characterized. The microbiome of the human gut, given its interaction with human health, also offers opportunities for synthetic biology. Synthetic biology has the potential to develop the next generation of probiotics. In addition, an emerging technology aims to use engineered bacteria in the detection and control of tumors. The idea is to engineer *E. coli* to sense the tumor microenvironment, which then leads to the invasion of the tumor by the bacteria, and the release of a cytotoxin to kill the tumor cells. Alternatively engineered bacteria might sense early cancers and produce a designed signal. It is important that biosafety of modified bacteria remains paramount. However, with increasing understanding of the gut microbiome the possibilities for early cancer detection are worth pursuing.

An important area where the application of synthetic biology can potentially make an important contribution is in cleaning up the planet. Products that historically have made important contributions to a particular phase of the development of human society have subsequently been shown to also have a downside. One area of particular concern is the plastic pollution that now covers the globe. Other areas are oil spills and soils from industrial sites that require remediation. The recycling of various plastics is gaining momentum, and there is a biological role in this reduction of plastic waste. An example comes from poly (ethylene terephthalate), known as PET. PET is an abundantly produced plastic used in packaging, bottles, and textiles. The bacterium *Ideonella sakaiensis* 201-f6, discovered in 2016, has this amazing ability to grow on PET as a source of energy and carbon. *I. sakaiensis* secretes a PET-digesting enzyme called PETase, which provides the monomers for its growth. Rational protein engineering has been used to produce bacterial *E. coli* mutants with improved efficiency of PETase. Contamination of water supplies with microplastics is an increasing problem. Engineered *E. coli* with enhanced PETase have shown promise in enhancing degradation of microplastic microfibers in wastewater sludge.

Enzymes are clearly very important in synthetic biology. Their capacity to act at the highest efficiency in the environment they are acting in is paramount. A breakthrough in this regard was in 1993 when synthetic biology was very much in its infancy. Frances Arnold, an American chemical engineer, together with coworker Kequin Chen, published a paper in the *Proceedings of the U.S. National Academy of Sciences* on what has become known as directed evolution. The alternative to the directed evolution approach is rational design, which requires a detailed understanding of the structure of the enzyme and its action. Directed evolution uses random changes in the amino acids of the enzyme, the tools of evolution—mutation and selection.

The first part of the directed evolution process involves introducing random mutations into the DNA of the gene. The mutagenized gene is transformed into a bacterium for synthesis of the randomly mutated enzyme from the DNA template. The enzyme is then tested to see which enzyme is most efficient in catalyzing the desired chemical reaction. The most effective enzyme(s) are then subjected to another cycle. In the 1993 Arnold study, there were three cycles of random mutation and testing. In this example, the natural enzyme, subtilisin E (a serine protease that digests proteins), was inactive in polar organic solvents. The directed evolution enabled the formation of an enzyme that could now efficiently catalyze in polar organic solvents, enabling its use in a wider range of chemical syntheses. This study established the principle that enabled maximum foreign enzyme activity when used in production pathways developed by synthetic biology. Arnold received the Nobel Prize for Chemistry in 2018 for her work on the directed evolution of enzymes.

Directed evolution has been used in the case of sitagliptin, an important drug for the treatment of diabetes. Sitagliptin is able to increase the secretion of insulin. Directed evolution in this case was not for an enzyme used for synthesis in *E. coli* or *S. cerevisiae* but in an organic chemistry process. The enzyme concerned was a transaminase that substantially improved the purity and efficiency of sitagliptin synthesis. Directed evolution can then enable the use of enzymes as manufacturing catalysts for the production of pharmaceuticals in a more demanding environment than that existing within the cell.

Directed evolution by introducing random mutations into the gene is one way of improving enzyme activity for specific conditions. There has been another more recent advance that can influence protein engineering. In nature, there are 20 amino acids used to make functional proteins. These amino acids have relatively simple functional groups that influence their biochemistry. It is now possible to have unnatural amino acids with a greater range of properties as part of a protein. How is this achieved? There are three codons that do not code for an amino acid, but when present act as stop codons. These codons have been given common names; they are UAA (ochre), UGA (opal), and UAG (amber). Given that there are three stop codons, this frees up the ability to change a stop codon to specify an unnatural amino acid. To understand how the unnatural amino acid is able to participate in protein synthesis, consider how protein synthesis works. The correct amino acid is attached to its cognate tRNA by an aminoacyl tRNA synthetase (aaRS). The tRNA has the correct anticodon to pair with the mRNA codon and deliver the correct amino acid. So, what is required for the use of the unnatural amino acid in protein synthesis is the construction of a tRNA that only delivers the novel amino acid when it base pairs to the mRNA codon (e.g., the amber codon). A novel aaRS needs to connect the unnatural amino acid to the novel tRNA. This difficult task of using unnatural amino acids can be achieved in *E. coli* by using unique aaRS/tRNA pairs derived naturally from bacteria such as the methanogen *Methanosarcina barkeri*. (The *M. barkeri* aaRS/tRNA gene pair uses the pyrrolysil-tRNA synthetase giving PylRS/tRNA**CUA**, which can be engineered for a given unnatural amino acid with the anticodon for amber.) Appropriate aaRS/tRNA pairs have also been developed for using unnatural amino acids in eukaryote cells.

The ability to incorporate unnatural amino acids with novel properties into proteins opens up many possibilities. It can be used to achieve increases in enzyme efficiency, but the greatest benefit is the ability to produce enzymes with no equivalence in nature—new to nature reactions. The ability to incorporate fluorescent unnatural amino acids into proteins offers a refined way to track the cellular activity of a given protein, without an add-on fluorescent tag. Unnatural amino acids are finding application in biomedicine in antibody–drug conjugates (ADCs). ADCs can be used to specifically target cancer cells with cytotoxic drugs. The antibody targets cancer cells with specific surface antigen markers to specifically introduce the cytotoxic drug. The use of unnatural amino acids to produce the conjugates improves the homogeneity of the product because of the ability to have more control of the conjugation site. Such ADCs have been used to target certain breast cancers that have specific markers on their cells. Unnatural amino acids are also being used in vaccine development by making viral replication dependent on unnatural amino acids, but not able to replicate in normal cells.

As synthetic biology advances, more complex gene circuits can be programmed into the cell. One of the major questions in biology concerns how multiple stable cell states can be produced during development from a single undifferentiated cell (Chap. 5). Genetically identical cell types are able to produce thousands of molecularly distinct mitotically heritable cell states. In this context, a synthetic gene circuit known as MultiFate has been developed. This was inspired by natural transcription networks that regulate stem cell differentiation.

The MultiFate circuit is built from artificial transcription factors. Each transcription factor has a zinc finger DNA binding domain and a dimerizing domain for the binding of another transcription factor. The transcription factors self-activate when two identical transcription factors (homodimers) combine but are inactive if the transcription factors are not identical (heterodimers). With MultiFate-2, with two transcription factors, three stable cell states can be obtained—AA, BB, and AABB. With MultiFate-3 by adding another transcription factor, up to seven stable cell states can be obtained with one, two, or three homodimers from—ABC, AB, AC, BC, A, B, C.

The MultiFate concept has been tested experimentally using cultured mammalian cells, with the expression of the transcription factors initiated by changes to the incubating media. In the case of MultiFate-3, three cassettes were transformed into the cells, each containing a synthetic transcription factor and a fluorescent-colored reporter. The colors of the fluorescent signal from the reporter, detected by a cell sorter, indicate which transcription factors are activated. The stability of each of the seven cell states was assessed by culture of each cell state individually. A given stable state was obtained for at least 18 days. The approach used in MultiFate can provide new insights into the way differentiation of stem cells and the development of multicellular organisms can proceed. It also provides a way forward in providing synthetic circuits for engineering therapeutic cell states and particular cell types where they may be required.

Many tools are now available for use in synthetic biology with promising applications in agriculture, industry, the environment, and biomedicine. Practical success has been achieved in a number of areas. The application of artificial intelligence to improve understanding of the regulation of biochemical pathways and enzyme structure and function will further advance synthetic biology. It is important that safety considerations be prioritized in applying the fruits of synthetic biology.

Further Reading

Arnold FH. Design by directed evolution. Acc Chem Res. 1998;31:125–31.

Chin JW. Expanding and reprogramming the genetic code. Nature. 2017;550:53–60.

Dou J, Bennett MR. Synthetic biology and the gut microbiome. Biotechnol J. 2018;13:e1700159.

Hale V, Keasling JD, Renninger N, Diagana TT. Microbially derived artemisinin: a biotechnology solution to the global problem of access to affordable antimalarial drugs. Am J Trop Med Hyg. 2007;77(Suppl 6):198–202.

Savile CK, Janey JM, Mundorff EC, Moore JC, Tam S, et al. Biocatalytic asymmetric synthesis of chiral amines from ketones applied to sitagliptin manufacture. Science. 2010;329:305–9.

Shao Y, Xue C, Liu W, Zuo S, Wei P. High-level secretory production of leghemoglobin in *Pichia pastoris* through enhanced globin expression and heme biosynthesis. Bioresour Technol. 2022;363:127884.

Son HF, Cho IJ, Joo S, Seo H, Sagong HY, et al. Rational protein engineering of thermo-stable PETase from *Ideonella sakaiensis* for highly efficient PET degradation. ACS Catal. 2019;9:3519–26.

Voigt CA. Synthetic biology 2020–2030: six commercially-available products that are changing our world. Nat Commun. 2020;11:6379.

Wen A, Havens KL, Bloch SE, Shah N, Higgins DA, et al. Enabling biological nitrogen fixation for cereal crops in fertilized fields. ACS Synth Biol. 2021;10:3264–77.

Zhu R, del Rio-Salgado JM, Garcia-Ojalvo J, Elowitz MB. Synthetic multistability in mammalian cells. Science. 2022;375:284.

15

We Have the Whole World in Our Hands

The ability to understand our environment, ourselves, and to be creative flows from the capacity of the human brain encoded in our DNA. It is remarkable that the universe and our planet, by providing the chemical building blocks of DNA, could then ultimately be perceived by us. In this chapter, I wanted to illustrate the paradoxes of humanity and consider our inherent capacities with all that has been achieved scientifically, and to an extent socially. Progress and prospects in combating climate change, and the DNA connection are considered. Contemporary attitudes are influenced by science skepticism by some groups; climate change and vaccination are illustrations of this. Human aggression and warfare in relation to the human species are commented on, including reference to a real-time study of tribal conflict and peace in Papua New Guinea. Gene sequencing in diagnostic medicine and AI for genomic analysis have important implications for the future. The scandal of gene editing of human embryos is illustrative of the need for regulation and close monitoring of gene therapy. Human fertility is considered by some as a significant future issue. Can the "super species" use DNA biology and technology wisely?

David Suzuki, geneticist and environmentalist, pointed out that we have been transformed from naked apes into a "superspecies." He argues, however, like others, that humankind has exploited the environment in a continued pursuit

R. J. Rose, *DNA Know Thyself*,
https://doi.org/10.1007/978-981-97-8381-6_15

of economic growth. Suzuki worries that a desire to conquer and control nature has led to this situation and that biotechnology is part of that. I would argue that our environment, our food, and our health can be augmented through our knowledge of DNA biology and wise application of that knowledge. DNA links us irrevocably to our environment, and both have the potential to sustainably flourish. The nature of living systems will always set up challenges as the COVID-19 pandemic has indicated. Our knowledge base helps us respond. In addition to sustaining life on earth, a challenge for humanity lies in the ability to live peacefully within and between societies. The human ecosystem can work for the common good, and altruism as a positive behavioral trait occurs, even given its links to its evolutionary role such as in kin selection, reciprocal altruism, or guilt-related altruism. While each one of us is unique, humans have evolved into a single species and in that sense we are one irrespective of where we live on the planet; DNA knowledge of the genetic code has emphasized that. Clearly taken as a whole, past, or contemporary human history is equivocal. In 2024, it is easy to be pessimistic given global warming, the COVID-19 epidemic, divisions within societies, and war in Europe and the Middle East. In this chapter, I wanted to illustrate the paradoxes of humanity and consider our inherent capacities with all that has been achieved scientifically, and to an extent socially. Can we save our species, or have we already peaked, peak humanity? Is it possible for humanity to summon its "better angels," or is there a downside to our evolution that is destructive?

Climate change provides a contemporary example of how humankind has difficulty in working cooperatively to overcome a serious problem. The continued increase in atmospheric CO_2 has been known for some time. Roger Revelle, a pioneer researcher on global warming, wrote in

1982 "that the atmospheric content of CO_2 is increasing seems well established." He pointed out that the link between atmospheric CO_2 and the surface temperature of the earth was independently posited in the late nineteenth century by Arrhenius and Chamberlin. The Paris Agreement on climate change indicated that all parties should strive to keep global temperatures "well below 2 °C above pre-industrial levels and to pursue efforts to limit the temperature increase to 1.5 °C above pre-industrial levels." Keeping the warming increase below 1.5 °C is unlikely to be achieved. A 2022 study published in *Nature* examined what temperatures could be expected if all conditional and unconditional pledges because of the 2021 United Nations Climate Change Conference (COP26) in Glasgow were implemented in full and on time. The results indicated that temperatures could be limited to 1.9–2.0 °C above pre-industrial levels. In some ways, this could be considered a positive given the difficulties in providing a type of international consensus. Given that we are currently living with a 1.2 °C above 1850–1900 levels and experiencing an intensification of heatwaves, drought, coral reef bleaching, and floods, there is an urgency. The hottest global temperature since records began in 1850 was recorded in 2023. This was 1.18 °C above-average twentieth-century levels[1] and 1.45 °C above 1850 levels.[2] The ongoing effects of COVID-19 and war in Europe and the Middle East have exacerbated the energy transition situation. It raises the more general question of the difficulty of engendering a sense of urgency. Current attitudes do appear to be consistent with the "Selfish Gene" argument as here individuals are concerned about individual security. Those who

[1] NOAA National Centers for Environmental Information, Monthly Global Climate Report for Annual 2023.

[2] World Meteorological Organisation.

continue to argue against climate change are frequently motivated by support for the status quo of maintaining CO_2 emissions, so as not to influence their livelihood. Those who support action on climate change could conceivably see advantages in new industries while others may be acting altruistically. Altruism here could potentially benefit the individual as well as the species. As Dawkins points out in his introduction to the 30th anniversary of the "Selfish Gene," our brains have evolved such that we can rebel against the selfish gene. The human intellect can hopefully reach a consensus on the most effective way to combat climate change and sustain a livable planet that will support all life forms. While it is possible that humanity will eventually vigorously action climate change, there is another aspect of our genes and human responses to global imperatives. There is some evidence from twin studies that that there is a correlation between genetics and political views, also making consensus more difficult.

Following on from the COP26 conference, there was a growing consensus on committing to action on climate change to have net-zero emissions by 2050. The evolving situation in Australia provides an example of the gradual move to combat climate change more seriously, even though consensus remains extremely difficult. Policy went back and forth for the 10 years prior to 2023 before there was support for more ambitious action on climate change; now net zero in 2050 has been legislated. While this is consistent with a rational understanding of a need to act, extreme events were likely more important in swaying minds. In 2019/2020, there were extreme bushfires (wildfires) and then in 2022/2023 extreme floods, supporting the predictions of extreme weather events due to increases in greenhouse gases leading to global warming. An unprecedented sequence of mass coral reef bleaching events occurred in

1998, 2002, 2016, 2017, 2020, 2022, and 2024. While the recent greater emphasis on combating climate change is encouraging, maintaining progress will require continued effort. Current global events, the ramifications of the COVID-19 pandemic, and disruption of energy supplies due to the invasion of Ukraine by Russia are stressing economies and slowing change, even given the extremes of climate in the 2023 Northern Hemisphere summer. It is, however, possible that increasing gas prices and using gas supply as a form of coercion will help in decarbonizing energy supply. Misinformation and disinformation in both some traditional media and social media outlets are a contemporary problem, requiring continual vigilance, in responding to global warning.

Increasing the capacity of wind and solar power, retiring coal-fired power stations, increasing electric vehicle sales, and methodically increasing energy efficiency are key for reducing carbon emissions. COP28 held in Dubai in 2023 agreed to transition away from fossil fuels. While this is encouraging in formally recognizing the significance of fossil fuels in causing global warming, it is a somewhat tepid outcome. It may be that technologies to remove carbon from the atmosphere (in addition to photosynthetic sources) may need to be developed. DNA biology and technology have a part to play (Chaps. 3 and 11). Maintaining thriving ecosystems and maximizing photosynthetic capacity are significant areas. Understanding micro and macro ecosystems is aided by DNA sequencing technology to understand all the organisms involved. This is particularly relevant to soil and ocean networks of microorganisms. Synthetic biology is being used in waste amelioration (Chap. 14). There is research underway to enhance CO_2 fixation in some plant groups as the photosynthetic metabolism of different plant groups are increasingly well understood (e.g., C_3 versus C_4 plants) and metabolic engineering can be applied using DNA technology. The ability of leguminous plants to use atmospheric nitrogen is also increasingly well understood. Understanding how this evolved has provided the possibility of engineering nitrogen fixation into non-leguminous plants like rice and wheat. In the final analysis, humankind must as a species respond to climate change.

An important contributing factor to the slow response to initiate combating climate change has been skepticism of the science by a substantive group. Another example comes from the uptake of vaccination in response to the COVID-19 pandemic. Many in the United States, a well-educated population with an extremely strong science base, still have a substantive group that have not received one or two vaccinations. According to the U.S. CDC (Centres for Disease Control and Prevention), 16.8% (5 + years) had not been vaccinated once as of July 2022. Further, 28.8% (5+ years) have not been fully vaccinated with two vaccinations. Writing as editor of the journal *Science* in 2021, H. Holden Thorp expressed the general pessimism scientists have felt in the pandemic. The editorial was entitled "Time to unfriend Facebook." "Do you unfriend Facebook or beat it at its own game." Clearly the world has changed, and scientists must recognize that being objective and articulating science with its nuances is not enough. Holden Thorp points out that how science is communicated, with its caveats, needs to be reassessed in the social media age. Trusted and "liked" science communicators who can engage on social media need to be present.

Despite the presence of what amounts to an anti-science voice, the development of mRNA vaccines has been an important positive example of DNA-related technology, and the science and the scientists involved have been celebrated by many. It is so powerful to see how the discovery of mRNA and the genetic code, the packaging and delivery of the mRNA have ultimately led to a vaccine that presents the spike protein to the immune system. The polymerase chain reaction (PCR), genome sequencing, and rapid antigen tests have become household words. The COVID-19 pandemic has not yet been completely beaten, but the world in 2024 is much more like 2019, prior to the pandemic.

The difficulty humankind faces in dealing with global events reaches the extreme with warfare between countries or different groups within a country; the inability to resolve issues without conflict. While a common theme in understanding human evolution has aggression and warfare as a hallmark of human nature, representing extreme competition, this neglects the human capacity for peaceful cooperation, empathy, and prosocial behavior. Human evolution must be seen in the diversity of social, cultural, archeological, and ecological contexts, as well as the contribution of the "Selfish Gene" concept. As pointed out by Johnathan Goodman in discussing the recent book by Jonathan Silvertown on the cooperative history of life, "selfishness can lead to co-operation."

Evolutionary biologists continue to make important contributions to an understanding of the evolution of human nature.

What is important, given the history of warfare, is not to forget the peacemakers and not to frame war as necessarily an inevitable consequence of evolution. Continued research on the details of how conflict leads to war could still provide insights into ways to reduce global conflict. How is it that powerful heads of some countries, often seemingly needlessly, lead their country to war, seeing conflict rather than reasoned resolution. Those Heads of State that advocate and lead into war also need to have a responsive population, again there is the question of how individuals obtain accurate information, with social media exacerbating this difficulty. On a more positive note, connectivity across the world now makes it harder to hide violent oppression.

There is an interesting study by Wiessner from the University of Utah and Pupu from Papua New Guinea that deals with war and peace in recent times in Papua New Guinea. This study provides insight relevant to small societies of the past that anthropologists have studied, but in addition gives some insight on how people turn from conflict to peace in real time. The Enga, the largest linguistic group of up to 500,000, fought more than 250 wars against other clans between 1991 and 2005 using guns rather than bows and arrows. By 2011, the Enga fought few wars. The population, tired of violence and village courts (distinct from Papua New Guinea state courts), awarded compensation to the victim's clan. All members of the offending clan contributed to the compensation; jailings were rare. The judgments sorted things out in a way that gained approval of all involved. However, the Enga arrangement was not long-lasting as their society continued to develop. In 2023, there was a major massacre (press reports of 150 deaths) in Enga province. It does suggest that as societies evolve novel structures to resolve differences need to evolve. In the international arena, such bodies as the United Nations, the World Health Organization, and the International Court of Justice are valuable. International bodies are important, and they need to continue to evolve in the light of research and experience, in the nuclear age. The United Nations' 17 sustainable development goals are important for our future.

Religion in practice has frequently caused war, not peace, with different faiths or subfaiths taking up arms against another. Religion again reflects a paradox, given that religions are predominantly based on or championed by individuals of peace. Theodosius Dobzhansky, an evolutionary biologist and geneticist, was also a believer in a personal God. He wrote in 1973: "I am a creationist and an evolutionist. Evolution is God's, or Nature's, method of Creation. Creation is not an event that happened in 4004 B.C.; it is a process that began some 10 billion years ago and is still under way." What is relevant in the DNA context of a religious belief is to also understand the significance of the DNA molecule as the universal genetic code, the diversity of life that extends from that and the unity of life. What is unequivocal is that both warfare and religion have been a part of human history. With pan-genomic data now readily obtained, it may well be possible to link genotypes to empathetic, cooperative phenotypes. However, I don't think this will help humanity in any direct sense. The important point is to coop the capacity that widely exists for cooperation and empathy.

DNA knowledge and next-generation sequencing are impacting diagnostic medicine and gene therapy continues to progress (Chap. 13) as do vaccination and antivirals for the prevention and treatment of virus infection (Chap. 9). Knowledge of complex traits of human disease and behavior may increase because of rapid sequencing opening the capacity to sequence many human genomes, it raises the question of how we use DNA technology with the toolkit that is increasingly available. The human track record is not great at using technology in only a positive way. The earlier chapters have focused on the progress made in many areas of DNA biology. However, rogue behavior has come about in recent times. Gene editing has been carried out on human embryos, but the Chinese scientist He Jiankui used gene-edited embryos to produce a live birth. According to the journal *Nature*, it was not done in secret and US scientists were aware of the work. Gene-edited twin girls were reportedly born in early November 2018. He Jiankui reported his work at a gene-editing summit on November 28, 2018. He claimed to have disabled a gene known as CCR5, which encodes a protein that allows HIV to enter cells. A mutation in that gene exists in nature, helping to protect from HIV infection. He was fired from his university in January 2019. In December 2019, He was convicted by a Chinese court for violating medical regulations by applying gene-editing technology to human-assisted reproduction. He received prison time and was released in 2022.

The He Jiankui episode highlighted the importance of the regulation of genetic modification in medical research, plus awareness that rogue operators are always a possibility. As indicated in Chap. 11, genetic modification in plants is highly regulated, arguably overregulated in some countries. In medicine, a chance to improve health outcomes by using whatever is possible is understandably strongly supported by those who have genetic disease and want to have children free of the disease. Gene editing of the germline is a different matter to the genetic manipulation of somatic cells. I suspect there will be tremendous pressure to allow germline gene editing in the years ahead. It is difficult to see any uptake of germline gene editing ending well. Focus should be on using genetic and biological knowledge to come up with strategies independent of germline manipulation. As pointed out by the CRISPR Noble winner Jennifer Doudna, in an editorial in the journal *Science*, there are many potential uses of genome editing involving somatic changes in individuals that are not heritable. Clinical trials using CRISPR are underway for patients with a range of diseases, including cancer, sickle cell diseases, and eye diseases. There will be continued progress in genetic knowledge and its application, but how it is used should continue to be closely regulated. The COVID-19 mRNA vaccines were rapidly developed and approved (as well as a DNA vaccine), illustrating how mainstream DNA-related technology can be accepted and the power of it.

The objective of humankind should be to live peacefully in harmony with the environment. Our knowledge should be directed at the health of all components of the environment and ourselves. In relation to our own species, there are concerns about reproductive health. R. John Aitken points to long-term changes in Total Fertility Rates due to "a combination of environmental, lifestyle and genetic factors" (also see Chap. 7). There is concern about a decline in sperm quality, decreased testosterone levels, and increases in testicular cancer in males. The increased use of assisted reproductive technologies (ARTs) encourages low-fertility genotypes in the population in the longer term. While all this is not an immediate problem in terms of world population numbers, it is a concerning phenomenon. The environment, while not the only factor influencing declining male fertility, is an important factor. Again, we see the need for the full appreciation of the connection between our environment and human existence. Knowledge of DNA biology and technology has the potential to enhance the understanding of ourselves and to enhance human life on earth if used wisely.

Further Reading

Aitken RJ. The changing tide of human fertility. Hum Reprod. 2022;37:629–38.

Cyranoski D. What's next for CRISPR babies? Nature. 2019;566:440–2.

Dawkins R. The selfish gene. 30th anniversary edition. Oxford: Oxford University Press; 2006.

De Waal FB. The antiquity of empathy. Science. 2012;336:874–6.

Dobzhansky T. Nothing in biology makes sense except in the light of evolution. Am Biol Teach. 1973;35:125–9.

Fuentes A. It's not all sex and violence: integrated anthropology and the role of cooperation and social complexity in human evolution. Am Anthropol. 2004;106:710–8.

Goodman JR. Why selfishness can lead to cooperation. Nature. 2024;628:260–1.

Meinshausen M, Lewis J, McGlade C, Gütschow J, Nicholls Z, et al. Realization of Paris agreement pledges may limit warming just below 2 ° C. Nature. 2022;604:304–9.

Revelle R. Carbon dioxide and world climate. Sci Am. 1982;247:35–43.

Wiessner P, Pupu N. Toward peace: foreign arms and indigenous institutions in a Papua New Guinea society. Science. 2012;337:1651–4.

GPSR Compliance

The European Union's (EU) General Product Safety Regulation (GPSR) is a set of rules that requires consumer products to be safe and our obligations to ensure this.

If you have any concerns about our products, you can contact us on ProductSafety@springernature.com

In case Publisher is established outside the EU, the EU authorized representative is:

Springer Nature Customer Service Center GmbH
Europaplatz 3
69115 Heidelberg, Germany

Batch number: 10683277

Printed by Printforce, the Netherlands